D0645257

About the Author

JOHN H. RICHARDSON is a writer-at-large for *Esquire* and the author of *In the Little World* and *The Viper's Club*. His fiction has appeared in the *Atlantic Monthly* and the *O. Henry Prize Stories* collection. He lives in Katonah, New York. He can be reached at johnhrichardson.com.

Also by John H. Richardson

In the Little World

MY FATHER THE SPY

●

AN INVESTIGATIVE MEMOIR

●

John H. Richardson

HARPER  PERENNIAL

NEW YORK ● LONDON ● TORONTO ● SYDNEY

HARPER ● PERENNIAL

FIRST HARPER PERENNIAL EDITION PUBLISHED 2006.

Designed by Nicola Ferguson

Library of Congress Cataloging-in-Publication Data is available upon request.

ISBN 0-06-051035-8
ISBN-10: 0-06-051036-6 (pbk.)
ISBN-13: 978-0-06-051036-7 (pbk.)

06 07 08 09 10 ❖ / RRD 10 9 8 7 6 5 4 3 2 1

for John H. Richardson
October 10, 1913–June 1, 1998

CONTENTS

And ye shall know the truth, and the truth
shall make you free.

—inscription in the lobby of CIA headquarters

MY FATHER
THE SPY

1

MEXICO, 1998

Mom calls. Dad's in the hospital, on oxygen. It's his heart.

I fly down. They live in Mexico in a big adobe house with cool tile floors and high ceilings. Servants move quietly through the rooms. Mom greets me at the door, telling me through tears that she found him last night flopped across the bed with his legs hanging off the edge. He lay there for an hour before he started calling her, then he apologized for bothering her. We both smile because it's just so *Dad*—he's always so polite, so maddeningly self-denying. Sometimes Mom cries out: "Don't ask me what *I* want! Just tell me what *you* want!"

I go into his room. With his dentures out and his head laid back on the pillow, he's like a cartoon of an old codger, lips sucked over his gums and grizzled chin jutting out. When he sees me, his face brightens.

In his pajama pocket he wears a handkerchief, neatly folded.

A few minutes later he gets up to go to the bathroom. I'm used to seeing him hobble around the house. He's been juggling congestive heart failure, osteoporosis, cirrhosis, and about half a dozen other major illnesses for almost a decade. But now the nurse takes one elbow and I take the other and he leans over so far he's actually hanging by his arms, bent in half with his chest nearly parallel to the floor. He goes three steps and pauses, rests against the bureau, then takes five more steps and rests again. Glancing sideways I see gray in his cheeks, a whitish gray, like dirty marble.

He makes us wait outside the bathroom. He won't be helped in there. So we stand in the hall and when the toilet flushes I open the door and see him shuffle to the sink. He leans down with his elbows against the yellow tiles and washes his hands. On his way out, he stops to put the toilet seat down.

My father was a spy, a high-ranking member of the CIA, one of those idealistic men who came out of World War II determined to save the world from tyranny. After hunting saboteurs and Nazis during World War II, after sending hundreds of men to death or prison camps during six years behind Soviet lines in occupied Vienna, after manipulating the governments of Greece and the Philippines, and the two terrible years when he helped depose the leader of Vietnam and stored up the raw material for a lifetime of regrets, he retired to Mexico and moved behind these ten-foot-high walls. His bitterness was the mystery of my childhood. Eventually I became a reporter and started trying to put his story together, but whenever I pulled out my tape recorder for a formal interview my father would begin by reminding me that he had taken an oath of silence. That was always the first thing he said: "You know, son, I took an oath of silence."

Later I started interviewing his old friends and colleagues, traveling to Washington and writing to Europe and New Zealand. Some were helpful and pleasant, painting pictures of a tough-minded, piano-playing spy who drank martinis till dawn and carried a gun through the ruined cities of post-war Europe—a man I could hardly imagine. But many of his friends resisted me. One refused even to have a cup of coffee. "I don't approve of what you're doing," he said.

"What am I doing?"

"You're trying to find out about your father."

Another time, I drove to Maryland for a meeting with a group of retired spies. But after the coffee and small talk, they started trying to discourage me. One said that my father would be angry if he knew I was asking questions. Another broke off in the middle of a harmless anecdote and refused to continue. The wife who refilled my cup told me that her kids never asked a single question. "I've had people ask me, 'What was it like being married to a spy?' I would say, 'Oh, was I married to a spy?'"

Tonight I set up my futon on the floor of the study, close enough to hear him if he needs help. Later he starts wheezing so hard I think he's about to die *right now*. The nurse pounds on his back until he recovers and a minute later he starts worrying again, this time about my mother and whether she's adequately covered by insurance and his pension, things we've gone over a million times before. He gives me advice on dealing with the house after he dies and tips on getting his estate through the Mexican system. I tell him not to worry, kissing his scabby forehead.

Back in the study, I crawl into my bed and take comfort in the familiar setting. This same furniture has gone with us from the hilltop mansion in Athens to the former secret police headquarters in Saigon: the red leather chairs my mother bought at a garage sale in Virginia, the capiz-shell lampshade from the Philippines, the drop-front desk and round cherry table Mom picked up in Vienna after the war, when gorgeous old furniture was selling for a *lieder*, the autographed pictures of the king and queen of Greece waving from red leather frames embossed with raised gold crowns, mementos of the glory years when Dad fished with the queen and squabbled with the foreign minister and ran spies into Bulgaria and Albania.

And the books—the books most of all. Bound in red leather with his initials pressed in gold leaf into the spine, complete sets of Aristotle and Plato and Cicero, the essays of Montaigne and the *Anatomy of Melancholy* and *The Confessions of St. Augustine*. There's plenty of George Orwell and Winston Churchill and volume after volume on communism, from *Conversations with Stalin* to the collected works of Lenin to more specialized titles like *Defeating Communist Insurgency*, but also *Leaves of Grass* and *Gargantua and Pantagruel* and hundreds of hardcover novels by Saul Bellow, Norman Mailer, Somerset Maugham, James Joyce, Lawrence Durrell. Sitting in one of the old red leather chairs, I've spent years dipping into them.

When I go into his room in the morning, Dad's asleep. He looks like one of those gaunt old men Rembrandt liked to draw. The room is small and quiet as a monk's cell. Next to his bed there's an arched window shaded by a large violet bougainvillea vine. I sit in a chair at the foot of the bed and read a book about the human body, research for a magazine article. After an hour, he wakes up.

"Good morning," I say. "How are you feeling?"

He smiles but doesn't answer. He's thinking about something else. "Do you feel guilty for stealing my name, Jocko?"

"Not really, Dad."

"I got that name in North Africa," he says, telling me how he got in a fight with another soldier and an officer told them to put on boxing gloves and take it into "the gladiator ring." He often speaks with this odd formality, as if he is translating not from another language but another time. "Neither one of us was very effectual. But after that they called me Jocko."

He doesn't say why. Perhaps it was a reference to some fighting Scotsman. But the memory sets him musing. "Accidents play such a large part in our lives. I don't mean accidents like car accidents. If it hadn't been for the war I would have had a very different life."

I've heard this a thousand times. Dad loves to extract lessons from things.

Then his tone changes, dropping a little. "Does Mike think this slump will get better?"

Mike is his doctor. For a moment, I'm not sure what to say. "I don't think so, Dad. There's always a chance but I don't think so."

He seems relieved. Behind his breath there's a rattle deep in his throat or deeper.

After lunch, his pulse plunges from eighty beats a minute to fifty and he starts gasping down lungfuls of air. He complains about pain in his chest and pain in his arm, then his face slackens and his lower lip sags back into his mouth. The nurse shows me the pulse rate dropping on his chart and walks over to the bed, taking his hand and stretching out his fingers. She rubs them softly, directing my attention to the fingertips. They have a blue tinge. She thinks she should call the doctor. I tell her to go ahead and a few minutes later my mother and sister gather by the bed, watching the nurse stroke his hands. With her round Indian face and solid little body, she has an inward calm that's soothing to us all. Taking her cue, I start to stroke him too, but I feel awkward and stick to his arm for a long time before I take his hand. Jennifer does the same on the other side of the bed.

I can tell she's feeling awkward too. We were the kind of family that never touched until we said goodbye and then gave each other quick hard hugs. But with the nurse to guide us, we caress our dying father as if we were normal.

Then he wakes up. "In the morning I'll try to take a bath," he says.

I can't help wondering if those will be his last words. "We're here with you," I tell him.

He smiles a sweet and grateful smile.

Then he remembers something. "Mike said there was salt in fish, but fruit is good," he tells me. "Remember that son, fruit is good for you."

We did not get along when I was a kid. He was distant and preoccupied and I was (I am told) a natural born smart-ass. By the time I turned fourteen I was sneaking out to take drugs, shoplift and commit acts of petty vandalism, which on at least one occasion prompted the intervention of the local constabulary. That was also the summer he told me he worked for the CIA. But I can't claim high political motives for my rebellion. The only possible connection is that in the summer of 1968, he was the kind of guy who would work for the CIA and I was the kind of guy who wanted to drop acid and listen to the *White Album* over and over. Late that summer we moved to Korea, where he brooded on the world's most rigid totalitarian state (just twenty-six miles north of our house!) and I dated Korean bar girls and smoked bushels of dope. Military intelligence officers wrote reports on my activities and sent them to my father, who gave me lectures on being a "representative of my country." This seemed pretty comical to me, since all my fellow representatives were just as whacked out as I was; Adrienne had a habit of carving on her arm with a razor, Karen was dabbling in heroin, and Peter dropped out of high school into a reefer haze. So I would bait my father at dinner by defending Communism until he got insanely angry, sputtering his way into lectures on totalitarianism before storming away from the table in disgust. Once I called him paranoid and he exploded into the most gratifyingly paranoid rage I have ever seen. It all came to a head the day I got beat up by an MP who was offended by my ducktail (he called me a girl, I gave him the finger). Dad came to bail me out with his chauffeur and big black car and forced me to apologize for making that poor MP beat me with his club. Not long after, the helpful men at military intelligence sent him a note saying I was a "known user of LSD." Then an Army psychiatrist had a crack at me. Soon I found myself on a plane back to the States. Sixteen years old and on my own. Thanks a lot, Dad. And fuck you very much.

That's how I felt at the time, anyway.

Today Dad is worried about the nurse. Has she eaten lunch? I tell him not to worry, Leonora's got it under control. But when she comes in, he asks her. "*Usted comió?*" She smiles and nods. He tells her to be sure and take care of herself too.

I help him to the bathroom. After the toilet flushes, I peep through the crack and see him bending down to wipe off the edge of the toilet bowl. When the *Titanic* sinks, he'll be the one putting down the toilet seat.

Back in the room, he apologizes again for taking so long. "I don't want to be a distraction."

Don't worry about it, I tell him.

Sometimes he can't understand what I'm saying and the conversation gets pretty surreal. He wants to hear the news and I tell him that yesterday they made peace in Ireland. He frowns. "You're in denial?"

"No Dad, peace in Ireland."

He still loves talking foreign policy. When I read him the news summary from *Slate* magazine, he says he likes Netanyahu and feels the Israelis can't take the risk of accepting a Palestinian state. He quotes Marcus Aurelius on the dangers of liberalizing autocracies and I tease him about taking stoic self-denial to the point of living on nothing but air and Clinton scandals, like some mutant Republican orchid. But his love of ideas always makes me a little sad. Especially now, I want him to have a more tangible kind of pleasure. So I get some ice cream and Jennifer spoons it into his mouth, accidentally dripping a few drops on his blue pajama shirt. He gets testy then and says he doesn't want any more, that she did it wrong.

Later we try again. "Do you think you could eat some Jell-O?"

He frowns. "Time to go?"

Now he's in the bathroom again. After the toilet flushes, I wait outside the door to help him back to bed. Finally I crack it open and sneak a look. He's frozen in the middle of the bathroom, staring at the basin in doubt and some alarm. It doesn't take long to figure out the problem. He wants to wash his hands but he's afraid to lean down. So I grab a washcloth and start wiping them, but then Leonora fills a basin with warm water and offers him the soap. She understands what he needs, which is to do it himself.

I hold out the towel.

Back in his room, I suggest he try to sit up for a while. Mike said it was better for his lungs.

He shakes his head. "Why do the so-called right things, when they'll just prolong this condition?"

He lies back on the bed, eyes closed, talking intermittently while I read the paper out loud. Half asleep, he mumbles: "It's Jimmy Hoffa."

Jennifer and I tease him about finally giving up the secrets.

He takes us seriously. "It has always been off limits for the Agency to conduct domestic operations," he says.

Tonight Dad's blood pressure shoots down from a hundred over sixty to eighty over fifty. He sees Jennifer in the hall and doesn't recognize her. "There's the lady who is going to give me my Metamucil," he says. But he still puts on his slippers every time he goes to the bathroom and still insists on having a handkerchief folded into the shirt pocket of his pajamas.

Lying back on the bed with his eyes closed, he asks me: "When did this happen and how? This condition?"

I don't know what to say.

He turns to my mother: "I'm sorry to be such a problem."

"You're not a problem to me," she answers.

"That's important," he says.

Sometimes he talks with his eyes closed and I can't tell if he's awake or dreaming. "Jenny went off with some guy, didn't she?"

Yes, Dad, she did.

"The CIA contact . . ."

Yes, Dad?

But he stops, drifting back into his dream.

One night he tells me a secret. For the last two days, he says, he's been hearing music—emotional music, orchestral, like a movie score.

In the morning, he says he's figured out where the music is coming from. "This music, it's produced by us. It's a subsidiary of ours."

Later he murmurs: "Yeah, this is the tail end." He looks at me. "I hope this never happens to you, to be partly killed."

Later still he frowns, puzzled: "This seems to be just a fragment of me," he says.

I'm in the kitchen when Jenny comes running through the door. "I think this is it," she says.

I run to Dad's room. He's squirming in pain, saying he wants a "Novocain" shot, wants to go to the hospital. So I call Mike and he warns me that if we take him to the hospital now, they'll never let him out. "It's a big business here. They'll hook him up to machines and keep him alive any way they can. It'll cost thirty grand and take weeks."

He doesn't spare the details. They'll open up his heart, put him on a ventilator, revive him repeatedly. He already told this to my mother, he says. We need to remember that Dad is an eighty-four-year-old man in stage-four heart disease. We need to be realistic.

So I go back and tell Mom the gist and she says, "We don't care about the money. I just want him not to suffer." So I go beyond the gist and explain about the open-heart surgery and the respirator and the stage-four heart disease and she changes her mind. We gather around Dad's bed and I tell him about the respirator and the heart surgery. He looks grave and nods his head. "Then we won't do it," he says.

We have to get him better drugs.

The next day, he suddenly gets much better. He takes a bath in the deep yellow tub and eats two eggs while watching CNN, sitting up for the first time in days.

A few days later, he's getting up to watch the news morning, noon and night.

I call the airport and schedule a flight home.

Before I leave, Dad says he wants to have a talk. "The primary concern is your mother."

Set your mind at ease, I tell him. "You've done a good job. You've left us in good shape financially. And Jennifer and I will take care of her from here on in. Just relax."

He thinks Mom's better off in Mexico and I agree, but at some point we will have to make practical decisions.

"Fine," he says.

"We'll just try to work it out as best we can."

"All the works of art and those two ivory carvings in the living room, I think you should be very careful about them in the sense that you don't just get robbed."

I will be careful, I promise.

"And the icons. They may be worth several thousand dollars."

"Okay, Dad."

"Be sure to get appraisals."

"I will."

"It wasn't very long ago that I noticed a diamond on your mother's hand."

"Mm-hmm."

"And I asked her how much it was worth. And she said the last assessment was twenty thousand dollars."

"Yes."

"Well, I mean if you didn't know that, you might settle for ten or something."

"I'll try to be very careful about all that, Dad."

"Okay," he says. "Okay."

Then there's a brief pause. "Did you remember to call the airport van?"

"I did, Dad."

"Well, give my love to Kathy and the two kids."

"I will, Dad."

"I apologize for being so useless," he says.

"It happens to everybody, Dad. It's nothing to be sorry about."

Then I read him the headlines: GOVERNMENT LOOKS FOR PROGRESS IN STATE REFORM TALKS . . . TIANANMEN SQUARE LEADER WINS FREEDOM IN THE U.S. . . . NETANYAHU OFFERS TO TRAVEL TO LONDON FOR PEACE TALKS. And a funny one: POLITICAL CORRUPTION FALLING FROM FAVOR IN MEXICO.

He doesn't see what's funny about it.

"You know, it's falling from favor. Meaning it was once in favor."

He waves my joke away. "Anything else?"

"You want me to read the Netanyahu piece?"

"Yes."

I read the piece. This is where we meet, journalist and spy joined in the hunt for useful knowledge. But the same impulse leads to so many clashes. Once, I even called the CIA public information office to ask for a look at his personnel records. A pleasant man named Dennis Klauer rang me back with the official response: "Not only no, but *hell* no—and if you pursue this, we must contact John Richardson Sr. and remind him of his secrecy oath." It was so frustrating, a clanging brass symbol of all the things I would never know in this life: my own father was classified top secret.

Two hours later I'm on a plane for New York, wondering if I'll ever see him again. Eventually I take consolation in something I got from him, an unexpected fruit of the same oath that divided us. Growing up as I did, I learned that people think secrets are a kind of magic. They think only mysteries reveal the truth. This is why they're so fascinated by intrigue, why they love codes and double-meanings, why they scratch at conspiracies with such anxious passion. For CIA employees, these are practically job requirements. But when you are a child in a house of secrets, you tend to find little glamour in mysteries. Secrets are just the papers on your father's desk. And politics isn't a conspiracy, and history isn't something you can decode with revelations from the vaults or theories about the *episteme*— they are something personal and intimate, a family romance written in headlines.

You gather the evidence and study the sequence of events, and slowly the patterns emerge.

2

A SPY IN HIS YOUTH

My father was born in Burma in a small village in the jungle south of Mandalay, where his father was drilling test wells for the Pan-American Oil Company. The year was 1913. For the first six months of his life they lived on the banks of the Irawaddy River, five miles on horseback from the nearest neighbor. Then his father came down with a rare tropical disease and loaded the family on the next ship back to California, settling first in Long Beach before moving fifty miles east to a small town named Whittier—the same town where Richard Nixon grew up, one grade ahead of my father all through high school.

In those days, Whittier was an orderly little place surrounded by orange and lemon groves. Founded by Quakers and named after an abolitionist poet, still tinged by their passion for modesty and social justice, it had forty churches and a ban against alcohol. My father's first friend was a Japanese boy who was good at making kites. He became a Boy Scout and devoured the adventure novels of Zane Grey and S.S. Van Dine. He studied piano but "never had any real feeling for music." He tried tennis, swimming, gymnastics, and chess and none of them clicked. He remembered a D in trigonometry, took Latin for three and a half years but "never reached Virgil," tried out for wrestling and track but wasn't any good at those either. It was a childhood, he always insisted, "average, normal, and quite undistinguished in any way."

Almost in passing, he mentioned playing basketball on a varsity team that won the Southern California championships three years in a row. But even then he seemed more interested in apologizing for being "known to many students whom I, in turn, did not know."

Shy around girls and self-conscious about the large strawberry birthmark on his neck, perhaps a bit disappointed at stopping a quarter inch short of five eleven, he was respectful by inclination. He taught Sunday school. He was a member of the Hi-Y Club, for students with "high standards of Christian character." He studied English and history and spent many afternoons at the town library, opening books at random, which got him interested in Nietzsche and Schopenhauer. He was best friends with a boy named George Chisler, another gentle idealist with a taste for books. When Will Durant's *The Story of Philosophy* came out, they devoured it together and spent many nights talking it over. Years later, Chisler told me that my father was the first person he knew who had ideas. "And they were his ideas, not something he got out of a book."

These details did not come easily. My father rarely spoke of his youth or his family, not even to my mother, so I had to badger him for letters and fill in the holes by tracking down long-lost friends and relatives. I was fourteen before he told me he had a brother. He wrote of his home life only once, when I was in college and hungry for any kind of contact, telling me his mother grew up on a poor cotton farm in Texas and his father might have graduated high school before starting work in the oil fields. On the ship back from Burma, a kindly surgeon saved his life. Then he drilled oil wells in Long Beach and almost made a million dollars once but got cheated on a handshake deal with a man named Rogers.

He was also vague on origins. He said his mother's folks came from Russia or Poland or Bohemia, his father's from Scotland by way of Ireland. They were Americans, that was the important thing. His father was a naturally intelligent man who read the newspaper carefully and kept up with both national and international events. An agnostic who had grown more liberal with the years, he had "an intense hatred for the Ku Klux Klan, which was riding high in those days. He hated the hooded men and he hated their brand, or any brand, of lynch or mob law."

His mother was also naturally intelligent, he said, though she never read anything and seemed to spend most of her time doing housework. Years later a cousin told me Annie "wasn't the kind of woman where you'd go crawl in her lap," but Dad always spoke of her warmly and put into that one college letter just a single hint of trouble:

> Both my father and mother were extremely dominant in their natures and had strong, often almost uncontrollable tempers. Neither was willing to give way so the family atmosphere was often filled with sound and fury. I remember that during one of their arguments, Dad jerked off the tablecloth and all the dishes from the breakfast table. My life-long reaction has been to avoid emotional violence and perhaps, to some degree, emotion itself.

Plunging on, apparently determined to get it all out in one burst, he wrote about the day his father died:

> He had come home, enjoyed a good meal, read the paper, and gone to bed. Without prior symptoms, the attack occurred and I saw him die in a matter of a few minutes, still remembering how his body heaved in a final spasm as mother rubbed the soles of his feet, trying vainly to stimulate circulation. I remember going out to the porch and praying vainly to God that he let Dad live. Negative.

Another thirty years went by before I found, in the small cache of secret treasures he kept his whole life, a picture of a black coffin with silver handles. In his careful hand, he'd written on the back his father's death date and also the burial date—March 29, 1928. On my fingers I counted down the years: he was fourteen.

So I add up the facts, move them around on the page. How did he go from the Boy Scouts to the CIA, from the Hi-Y Club to guerrilla war? With the income from two modest oil wells and his mother's nursing job, he and his brother were left relatively comfortable. But the Depression shocked them. Then Annie got married again, another dark memory he almost never discussed. He despised his stepfather and began spending as much time as he could fishing and camping in the Sierra Nevadas, often in the company of his cousin Ray and Uncle Roy—the only man besides Chisler he remembered often and fondly, frequently mentioning his Cherokee blood and great skill at tying fishing flies. Then he plunged into classes at Whittier College, a Quaker-inspired school where smoking and drinking were forbidden and the penalty for sex was expulsion. At first, he made a stab at normal college life. Following in Nixon's footsteps, he joined a sportsy social club called the Orthogonians. He tried out for the basketball team. Then, abruptly, he decided to give up sports to study religion. He always said he was too short for basketball, although I suspect the gloom sparked by his father's death and hated stepfather played a role. But after a few months of preparing for life as a Baptist preacher, he stumbled across a book called *The Varieties of Religious Experience* and decided that traditional religion was conventional and dogmatic, the dried-out shell of true spirituality, that true prophets experienced faith "not as a dull habit but as an acute fever." This launched him on a study of a Quaker philosopher named Rufus Jones, a passionate social activist who promoted something called "affirmative mysticism"—by blowing on the divine spark within, a young man could grow a holy flame that would be a light unto the world.

So my father tried blowing on his divine spark, joining the ranks of sensitive young men who listened to classical music and wandered the hills, reading poetry under oak trees. Later he remembered one of his friends mooning around the campus in a daze, "completely bemused and deballed by Walt Whitman."

Then he took another lurching turn, deciding that mystics were like "violinists so entranced by their own music they convinced themselves it came from the spheres." Inspired by the head of the English department, a man named Dr. Albert Upton, he gave up religion and turned so obsessively to the study of literature that by the middle of his junior year, Upton approached him as he sat reading under a tree. "Enjoy the simple pleasures," he said. "Don't become a 'greasy grind.'"

But he kept on studying and studying, haunting the library and withdrawing from friendships and college activities, until finally a day came when he sat in class and opened his mouth and nothing came out—just a hoarse croak and then silence. The moment stretched out and the class waited and a feeling of panic seized him.

They called it overwork. Suddenly he couldn't go to class, couldn't feed himself, couldn't even get out of bed. His friends had to rally around to care for him. When he came out of it, he knew that everything was different, that he was going to have to leave this small town and begin some great transformation. So he took a first step, transferring to the University of California at Berkeley.

This began his romantic phase. Finding himself completely unknown in a large city, submerged among fifteen thousand students, he moved into an off campus apartment and tried to float out of his gloom on a wave of beauty, gazing across the San Francisco bay and taking dreamy walks through the eucalyptus groves. He immersed himself in Shelley and Byron. He wore a corduroy jacket and a "flowing, multi-colored tie." He fell into a Quaker study group that set his Whittier ideals in a new light. Since Quakers believe everyone is equal and that you have to act on your beliefs, doesn't that mean taking a political stand on the great issues of the day? Doesn't that mean sticking up for the working man?

It all came to a head during a rally for the Abraham Lincoln Brigade, then warming up for the Spanish Civil War. He got so fired up that he marched into the Communist Party headquarters and approached a recruiter, a story he always enjoyed telling in later life because it gave him a sterling opportunity for self-deprecation. Once, I managed to tape-record him.

"Thank God, the recruiter rejected me out of hand," he remembered, "possibly thinking me a provocateur, or at least clearly recognizing me as a callow youth not knowing what he was about."

What did he say, Dad?

"'Go away and think about it some more.'"

T hat is his story as he told it when I was in college, garnished with a few details I scrounged up over the years. But from the beginning, I was certain that something was missing. It was implicit in the lessons he tried to pass on, in the way he swaddled his pain in abstractions. Here's how he resolved the turmoil of his breakdown in one typical letter:

> I found myself impressed by a key, epoch-making book on Romanticism and Classicism, Irving Babbitt's Rousseau and Romanticism. It is very much worth your reading.

Twenty years later, I go down to Mexico and my mother gives me a box of old letters in yellowed envelopes. "I thought you'd be interested in these," she says.

For the ten pivotal years through college and World War II, I discover, my father had poured out his deepest feelings in hundreds of letters, drawing a generous map of his journey from breakdown to self-mastery. Most were written to George Chisler, others transferred to him for safekeeping. A few years ago, Chisler came down to Guadalajara for a visit and gave them back. They'd been sitting in the bottom of a file cabinet ever since.

The first envelope is postmarked October 11, 1934, the day after my father's twenty-first birthday. He has just arrived at Berkeley and although it's a shock at first—his ideals of Quaker simplicity clashing with the "false sophistication" of the big city—he quickly finds comfort in the bustle of the big world. In a single evening at the International House, for example, he meets an English girl, an African man from the Gold Coast and an American boy from China. The campus is beautiful, the professors marvelous. "They have succeeded in reviving my faith in culture," he says, "a faith which must stand between me and utter despair of life."

Scanning down, I see sentence after sentence blooming into desperate Romantic proclamation. "Without intellectual and imaginative interests I should be absolutely dead . . . I should rather be a half-crazy poet than a healthy, stolid, well-balanced animal." Even the signature is different. With a flourish, this man signs himself *Jack*.

But the revelation is the footnote:

> George, I want you to know that Dr. Upton was the best friend I ever had on the faculty of Whittier College. At every point he tried to keep me from making mistakes, but I wouldn't listen to him. He warned me against eternal study and too many books. He warned against being a "library hound." And long before my trouble he warned me that I should be unable to mix with society if I continued in the monastic anti-social course which I followed for so long. He did his best to prevent my following the mistaken course which led me into much unhappiness. I shall be eternally grateful to him (and to you).

He was talking about the breakdown, but fresh, without the distance of time and abstraction to muffle the pain. With mounting excitement, I flip to the next page. The next letter is postmarked in mid-January, a few weeks after Jack returns from Christmas vacation in Whittier:

> I write to you from Olympus where gods hold forth in rolling tones, worthy of our ignorant admiration. But of course even these, seeming gods to me, are simply mead

bringers, servants to greater artists, creative transmitters of created beauty. And as for me, well, I'm the kitchen rat, damn it and damn my crude, frontier heritage devoid of Latin, Greek, Italian, French, German, and decent high-school educators. And damn the theory of play, play, play at the expense of so much that is rich in life. Our greatest mania is sport and more sport. Well, I say—to the hot place with so much sport!

Who the hell *is* this guy? He sounds more like me—an old-fashioned 1930s sophomore version of me—than my reserved and distinguished old man. And this boisterous stranger tells Chisler that gradually, he's been pulling himself up out of that period of frustration and now he's determined to change his life. He cannot settle for an ordinary life and an ordinary world:

> Most of us are satisfied with too little, and we never live even though we think that we do. We're pigmies, we're all the hateful, disgusting things that Swift said that we were, and the damnable thing about it all is that we seem complacently, oilily content about the whole matter. Well, I have no intention of resigning myself "to sleep alway." By the Lord, I'll escape this pigmy state if I have to spend the rest of my life doing it!

And on he rolls, the manic tone building, ambition piling on ambition, until finally he takes the edge off the intensity by confessing that he's found a sweetheart: Katharine Hepburn, who appeared that year in "one of the greatest works of art ever filmed," a movie called *The Little Minister*. Hepburn combines intelligence with wildness, refinement with the unconventional life of nature. She isn't like those Whittier professors who wasted their lives studying beauty only to succumb "to the taming tendencies of civilization." She's a gypsy, a wild pagan beauty-sprite dancing in the woods! *That's* the way to be, George!

> And when I compare the girls around me to her, they all suffer terribly; they seem too accursedly shallow, pale and colorless. Damned shallowness! Damned conventionality! We

should all cast off our conventional clothes, our dusty black and
grey, and dance the gypsy dance, clad in the gypsy garb—

Chisler must have written back in alarm, because the next letter Jack
posts seems designed to calm him down. "Your writing sent me off on an
entirely new course of reforms," he announces. Yes, he'd been strutting
about on the mountaintops, but now he was down in the shadows.

Candy, cigarettes, indifference, enervation, lack of confidence—all these small things gradually tore me down from
the heights on which I advocated wildness and untameability.

And baldness. Baldness is not to be discounted, George. He'd always
maintained that a man should have—along with a car, a radio, a home, and
a wife—a full head of flowing locks, preferably something that summoned
images of Vikings. Instead his white skull is rising through his scraggly
tufts of hair "like some great continental block."

Then there's this persistent sense of ambiguity. One minute he's soar-
ing over the rooftops and the next he's down in the slime looking for something solid to save him. One minute he goes to see a Phillip Barry play and
splits himself laughing at the line about people who have "an exaggerated
sense of their own unimportance." And a second later he's glaring at himself in the mirror convinced that Swift was right, that if we saw ourselves as
we really are, we'd hate ourselves.

For I find in myself everything that's ugly: intellectual
pride, intellectual groveling, self-hatred, despair, selfish hope,
scorn for little personalities, and then discerning unexpected
depth in men where I thought there was only shallowness and
discovering unexpected weakness in myself—and supreme
among all these forces is the hatred of the triviality of life,
combined with the feeling that it could be grand.

A month later, he's writing again with a breezy tone and news of a
completely different kind. Yesterday at Berkeley they had a "great student
strike." As the crowd spread across both sides of the street and deep into
the campus, the president of the local Humanist Society and a representative from a laborer's union joined the student leaders on the podium.

Except for an unthinking simpleton who tried to liven things up a little by cutting the loudspeaker cable, the whole event was "quiet, peaceful, and orderly"—a point he makes repeatedly, concerned about the propriety of this protest business.

> All in all, it was an inspiring scene. Most of the group seemed to mean what they said, and there was a general air of sincerity and earnestness. The crowd passed resolutions against compulsory R.O.T.C. and against future war.

In other letters, Jack and his friends toss around phrases like "the class struggle" and "the common man" and worry about their responsibility to organized labor, dismissing Republicans as reactionary by definition. It seems that Jack has landed at Berkeley smack in the middle of an early wave of American campus radicalism, a time when students were discovering issues like solidarity with labor, civil rights, women's rights and even free love. Jack's Quaker study group has counterparts all across the country with names like the Current Problems Club and the Pen and Hammer Club, while the Hearst papers fulminate daily against "red professors" who teach Communism and anarchy and colleges from UCLA to CCNY are suspending and even expelling students for protesting against Fascism and war. It's all so eerily reminiscent of the 1960s, I'm stunned my father never mentioned it. Like the flowing tie and his taste for Byron, it might have given us something to discuss.

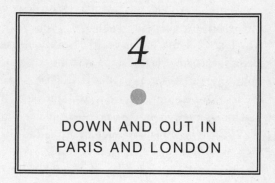

4

DOWN AND OUT IN
PARIS AND LONDON

In the letters he wrote me when I was in college, my father skimmed over the next two years in the style of an educational travelogue: After getting his bachelor's degree from Berkeley in the spring of 1935, he went to Paris, where he briefly attended the Sorbonne and found it so boring that for the first time in his life, he fell asleep during lectures. He went to art galleries and took a bicycle trip through the Loire Valley and spent a few nights in haystacks—not as romantic as it sounds, son—then got wildly drunk with a Scotsman named Ed Caird and left with a splitting hangover for Berlin, where he lived and studied German for eleven months in a *komaradschafthaus* with sixteen young Germans. He read *Mein Kampf* and saw that "malignant national cheerleader" give one of his famous speeches, but found it so difficult to argue with the articulate young Germans that he began to think there was some merit in the National Socialist system, a moment of intellectual weakness he came to understand as a vital lesson in the power of propaganda. Then he went on to London and bought a five-dollar bicycle, setting off on an epic trip through the Lake District all the way to Scotland, stopping frequently along the way to visit the homes of famous writers. In his letters to me, this is a big theme:

> I don't think Carlyle is read much these days but he was during my time. This Scottish writer was famous for his French Revolution and for Heroes and Hero Worship, in which he expounded the great man theory of history, namely that great men do have a significant influence on the course of events.

With a few more quick sentences, he describes crossing to Ireland and riding past the Lakes of Killarney, stopping to kiss the Blarney stone before pedalling back to London and the Salvation Army. And with that his wandering years are over. "The trip gave me the experience of seeing personally so much that I had read about in English literature and history," he concludes. Now it's time to give up literature and study something that might teach him more about the world—sociology or maybe anthropology.

If I had been reading carefully at the time, trying to make sense of the sequence, I might have stopped there. Wait a minute, pal. You're so obsessed with books that you *sleep in fields* to bike from one literary site to another and then—poof!—you decide to give up literature?

Now consider the way he told the story to George Chisler—the way young Jack told the story. He starts his journey out of Whittier on the Greyhound bus to New York (forty dollars), where he goes to Coney Island and the Follies but skips the Metropolitan Museum, resolving to tour it thoroughly on his next visit. Then he takes a ship straight to Paris (sixty dollars) and the Latin Quarter, the fabled Left Bank neighborhood where so many writers and artists have sipped absinthe and contemplated demoiselles, dropping his bags at a cubicle in a big warehouse room at the Salvation Army (twenty-five cents a day) before setting out to explore the city. "However, it wasn't nearly so romantic as I fancied, and after a few days of extreme loneliness and homesickness, I went to the Fondation des États-Unis."

This is the American cultural center, where he lingers over the free international newspapers for a few weeks until the director of the place offers him a room and a job, cashier in the center's restaurant. There Jack stays until he finds work at a 17th-century château owned by "an American gentleman of somewhat shattered fortunes," which accounts for the exotic letterhead he writes under, Château de Gregy in rich black letters. But it seems an odd kind of job.

> I was especially delighted by having breakfast in my room and by taking tea and cookies at five o'clock every day. I began to feel truly aristocratic and—whisper it sorrowfully—I almost deserted all my socialistic sympathies.

He never does tell Chisler what the job entailed, which was cataloguing this old gentleman's vast pornographic library. Then he comes back early from a short trip and surprises the old gentleman entertaining a male visitor, a social error that brings his château idyll to an abrupt and embarrassing end. So he buys another bicycle and sets off on a three-week tour through "various historical sections of France," from Chartres to Mont-Saint-Michel through Tours and Orléans to Fontainebleau, studying naves and nevures and flying buttresses and committing to memory the vital differences between Louis XIV and Louis XV. Back in Paris, he spends a month living on a dollar a day and waiting for winter classes to start at the Sorbonne, where he plans to study as much history, literature and art as he possibly can. As he explains to Chisler, he's still hungry for the feeling of visionary beauty and fullness that only learning can give a man. "Paris doesn't mean a damn thing unless you can appreciate the Gothic architecture at Notre Dame."

Then everything changes. After three months at the Sorbonne, Jack posts a Christmas letter that sounds downright breezy. Life has become tranquil and commonplace, he reports, and he's finally beginning "to lose the illusion that a fellow can find glorious, fairy-like grand things by going to far-off places." After a fellow has seen all the paintings and statues and châteaus, he's still pretty much the same fellow. Gazing upon the glories of Fontainebleau doesn't make him marvelous or exceptional or give him license to look down on the unrefined bumpkins back home. Paris is still interesting and all, but it doesn't make him "slop over emotionally."

Part of the change seems to be a social life. Living again at the Fondation des États-Unis, he's fallen in with a group of guys he calls the Anglo-American Club, four Brits and three Yanks who play ferocious ping-pong games in the lounge and gather in his room to pitch francs, loser buys the beer. Then they head down to the little bar on the corner and talk till the small hours—and for the first time in his life, he doesn't regret spending time away from his books. What a miser he was about his time! What an irritating spectacle he must have made!

> Only now do I realize how much I missed during those hours of withdrawn, alone, egoistic study. Strange that I should have been so tiny a pedant without realizing it.

Even travel has paled.

> I feel more and more that where a man's been doesn't amount to very much—not very much. All that which means anything is down there in the complexities of his own soul.

Ditto the glory of the Sorbonne. Sure, it has some of the greatest professors in the world, but that doesn't seem so important when you actually have to sit there and listen to them hour after hour. Fact is, he's going to classes mostly to improve his French, and Whittier and Berkeley are looking better and better. "So you see, George, I'm not becoming a learned man at all, but I am having a good time; I'm living like a human being."

And what do they talk about in their Left Bank *cafés*? The war in Abyssinia, for one thing. Mussolini invaded in October, using bombs and poison gas against helpless Africans in a shameless attempt to expand his Fascist empire. The French socialists and radicals started going crazy with protests and then the brand-new French Prime Minister—a Fascist sympathizer named Pierre Laval—made things worse with a peace proposal to Mussolini. And Jack's siding with the radicals. Italy was wrong to invade and the bombing of the hospital at Dessie was outrageous. He thinks Roosevelt is handling it all beautifully, especially in his refusal to bend to the business lobby and the chamber of commerce types who saw a chance to "make profits out of human slaughter."

But the real clue to his new mood is a short note he tucks into the envelope for Chisler's sister. Apologizing for being so serious in the main letter, he asks her to indulge him as he romps a little, taking as his theme the creature he might be when he finally returns:

> A bright-eyed little elf, perching myself on one of your sofa pillows? I'd have a voice like the high notes of a violin,

and I'd tell you all about the tiny, malicious things I'd seen. Or more likely, squatting on that pillow like an old toad, with broad nose, barren-topped head, wrinkled ugly skin? Or perhaps I'll come riding by your mansion astride a great warhorse, and I'll clank in the house majestic and powerful in gleaming armour, then tell you about great glorious lives in far-off lands.

But suddenly he breaks it off, starting a fresh paragraph: "You know, Virginia, I'm really too old for this bombastic nonsense. As a matter of fact, I solemnly swear that I will never do it again."

And he never did.

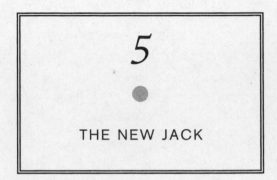

5

THE NEW JACK

So winter turns to spring and now Mussolini's troops are halfway to Addis Abbaba, and the French public is starting to cheer them on. When a law professor at the Sorbonne calls for sanctions, his students mount a protest against him. Jack watches from the sidewalk, standing in the rain and waiting for something to happen. But they just give speeches and call each other names.

Lately he's been reading about America, Lippman's *The New Freedom* and C. A. Beard's *The Open Door at Home* and also *Time* and *The New Republic*. He's been pondering Roosevelt and Al Smith and the Liberty League and the American Legion and the Reciprocal Tariffs and the Neutrality Act. And he's been making a study of his friends, who now include a young Frenchman working on his graduate degree at the Sorbonne. All day and night, this poor student walls himself up in his room, poring over books to no good effect. There were others like him, good fellows who simply lacked the talent to do what they were trying to do. So they became all bottled up in a futile intellectual struggle that only intensified their unhappiness.

Good Lord, anything would be better than that—even manual labor!

He's even starting to sour on Katharine Hepburn. *Break of Hearts* and *Alice Adams* were absolutely foul and he isn't sure he's even going to bother to see *Sylvia Scarlett*: "She's played only one type of role since she began, and that business of always holding her mouth open is a rather mechanical trait."

So instead of finishing his Sorbonne year, Jack rushes off to Berlin right after Christmas. In February he sends Chisler nineteen closely written pages that open the window on his soul, the first and last glimpse of what drove him away from his home town and up to Berkeley. Yes, he admits, he has been keeping his real feelings out of his letters. The silence is part of his inner revolution, his ongoing rejection of the "greasy grind" who snubbed sports and avoided his friends, who grew careless about his appearance and refused to take on any mundane task that might interfere with his intellectual development.

> To outsiders my actions must have seemed unworthy of a healthy man. And all the time I had the disturbing feeling that I was not meeting at Whittier College tests as difficult as other people of my own years were meeting at large universities.

Eventually, he developed the noxious Romantic idea that the bravest thing he could do was to "risk the exposition" of his individual truth even if that exposition "involved the description of intimate personal experiences." In the name of truth, therefore, he revealed many things about which he probably should have kept silent. He attacked sports, Christian fundamentalism, lethargic professors, all his opinions and feelings *spread out for anyone to see*. But he couldn't help noticing that his friends were beginning to feel superior to him, that in the wisdom of their silence they looked down on him and thought him egotistical to talk so much. Perhaps they even thought him *weak or unstable*.

> The antagonism which my course of life seemed to have aroused among my acquaintances is probably one of the factors which has ended up making me wish that other people know nothing about my own affairs. It seemed that when other people knew of your problems or ideas, it only provided

them with subject matter for ridicule, laughter, or gossip. The person whose inner nature was too open seemed weak and naive and child-like.

Then came the day George Ulner and Gerry Polenty accused him of intellectual dishonesty. That hurt, George, it hurt very much, especially because he thought he was doing his best to be honest. Certainly it played a part in his move to caution and silence. But the real problem with living emotionally is that emotion is so *inconsistent*. He's been a mystic one moment, a cynic the next, sympathetic to capitalism in the morning and by nightfall fired up with "radical socialistic hopes." He goes from Stoic to sensualist, from Shelley to Maupassant. And deep down, maybe all of his thoughts and opinions are nothing more than reactions to the environment. Maybe they're hints of an indecisive nature.

Lately he's been haunted by a passage in Aldous Huxley's new novel, *Point Counter Point*:

> Better to cultivate his own particular garden for all it was worth. Better to remain rigidly and loyally oneself. Oneself? But this question of identity was precisely one of Philip's chronic problems. It was so easy for him to be almost anybody . . .

At this point, Jack is five pages into his endless letter. I picture him writing it on a table in some Berlin beergarden, propping his copy of *Point Counter Point* on an ashtray as he sends his neat slanting letters in an orderly march across the page. Then he takes a deep breath and plunges on. The truth is, the Sorbonne is much more difficult than American schools. French kids learn Latin and Greek and philosophy and literature before they get out of high school, and the professors don't supervise you at all, you can go to class or not, depending on your whim. The result is a permanent subculture of layabouts who call themselves students but do no real work beyond eating, drinking, and going to bed with each other, a kind of soft intellectual degeneracy he cannot respect. This is one of the reasons he left for Germany, where he feels much more comfortable. He likes these students a thousand times better.

> Most of them are big, husky, friendly, good-humored, slow fellows; and they are well built. Remember, George, this is

confidential. I don't mean to get in bad with Frenchmen, and my judgment is probably unfair, but the French students need more ditch-digging or more good sports. I'll never rail against too much sport again—better that than too much intellectual life!

Fact is, George, a fellow can go to seed on all this culture business if he takes it too far. A man has to *act*.

Something else has changed too. His nature has become more sensual, his desire for "the female of the species" stronger than ever. Maybe it started in Paris, where the women dressed in a most seductive manner and he spent so much time in museums filled with paintings and statues of naked women. He still doesn't want to take on the drudgery of family life, doesn't want to get bogged down with some bridge-playing housewife, still wants a girl who is intellectually awake and understanding and kind and gentle and wise and strong and courageous, someone who reads and goes to plays, who has definite life concepts and knows how to defend them, who leads a physical life and stays slim. Not that a slug like him would have any right to a girl like that. But now there were times when the physical need was so strong he'd take any stupid female so long as she was fairly pretty.

So yes, darn it, he would like to have a woman.

But now he really has to stop writing. The fellows at the *komarad-schafthaus* are standing around making cracks at him in pidgin English. They're good friends, he says, the best-humored bunch of fellows he's seen in Europe. But he'll write about that another time, in letters that are more moderate and healthy.

And one last thing:

> Remember what I said about five pages back about this letter being only to you, old man. I've said some things I'd not want anyone but you or some other close friend to hear.

Already, the habit of secrecy is taking hold.

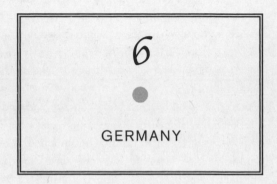

6

GERMANY

Five months later, Mussolini annexes Abyssinia and Franco begins the Spanish Civil War and Jack is back in London, dutifully trudging past miles and miles of Gainsboroughs and Hogarths and depressed because he misses his "German friend"—someone explained in a lost letter, another missing piece of the puzzle. But now a new confusion plagues him. Germany was so clean and friendly, so neat and shining and bourgeois and staid. He liked the food, the beer, the pastry, the people. He especially liked their frank earthiness with regard to sex. In fact, he was sure that D.H. Lawrence got inspiration in Germany for *Lady Chatterley's Lover*, which was still banned in England and America. But Germany had a more problematic side. After reflecting on the matter for some time, he decided not to say *heil Hitler* as a form of greeting. It seemed utterly unfair to require someone to take a political loyalty test with every hello. So he didn't say it and in fairness to his friends at the *komaradschafthaus*, they carefully avoided putting him in an embarrassing situation. But to be honest, if he were a German refusing to salute, they might not have been so considerate.

And Germany was so very *controlled*. All the newspapers said the same thing. Many books and films were *verboten*, like the work of H. G. Wells and *All Quiet on the Western Front*. Hitler encouraged a fanatic nationalism that constantly emphasized the antagonism of the rest of the world, making people feel even more embattled and patriotic.

> Most German citizens know nothing or almost nothing about concentration camps, secret police, university corruption, brutality toward Jews and communists, etc.—all that is carefully suppressed and the average German citizen seems as ignorant as a baby concerning these subjects.

And make no mistake, Hitler's line about wanting "peace with honor" pretty much justified anything he wanted to do. "I trust Hitler's 'peace' as far as I can throw a Texas steer," he says.

So off he sets on his bicycle trip around the British Isles, giving Chisler the same vaguely abstracted travelogue he later gave me, complete with the Lakes of Killarney and the home of Thomas Carlyle. But in this version of the story, the British Isles just make him miss Germany. And God knows they had a long and bloody history of political oppression. So why not ignore politics and trust his feelings? Why not go back where he's really happy?

He stops in London just long enough to pick up his mail.

But when he gets back to Berlin, his confusion hits him with fresh force. Writing to Chisler a month after his arrival, he speaks of almost nothing but the German paradox. This is probably the best-educated country in Europe, he says, where the chief entertainment is conversation—a fellow sits at a café with friends and coffee and cigarettes and drinks and talks and talks and talks. Just to meet the conversational demands, you're compelled to keep up with cultural and political events. And even though Paris is more beautiful and London is more interesting, he just feels more *comfortable* in Berlin.

But there are so many troubling details. In London, the headlines said U.S. WINS OLYMPIC GAMES. In Berlin, everyone insists that the Germans won. And one day he's talking to a perfectly nice young Austrian who tells him that back in 1933 just after the Nazis came to power, a friend of his got excited and cut off an old Jew's beard and was immediately deported. This proves the courts aren't unfair and Germany isn't really anti-Semitic, the Austrian insists.

And maybe there is *some* truth in it.

> I have never seen anyone using violence on Jewish people, nor have I ever spoken with anyone who has been a witness to such violence. I do not say that violence has not occurred. I only say that I have not come into contact with it in any way. I do admit that violent remarks about Jews are published in newspapers, and that stern legal measures have been taken against them. A few days ago I went to the largest community pool in Berlin for a swim. On the door there was a sign: Jews not wanted. I am told that such signs are also posted on the outskirts of many villages.

What to think? What to do?

> Before you become to indignant, George, remember that we lynch negroes in our country, that we have had the Ku Klux Klan, that we have capitalists who have used machine guns against strikers, factory owners have employed strike-breakers ten percent of whom have criminal records, and that tarring and feathering are not obsolete practices even in our land of freedom and justice. There is no reason to *hate* Germany for the perhaps mistaken and cruel policy which she is pursuing against the Jews. We may strongly disapprove of her practices, but why should we hate Germany for these things any more than we should hate our own country for the cruelties and injustices that daily take place there?

After another six months, in March 1937, Jack sees Hitler give a speech denouncing the Treaty of Versailles and comes to the conclusion that this time Hitler has a point: after all, Germany only signed the "sole guilt" admission under great pressure, and the leading historians all seem to agree that Germany was not the only nation responsible for the First World War. "Therefore," he tells Chisler, "I admire this particular action of Hitler's."

> Please don't think I've become pro-Hitler or pro-National
> Socialism. I want to remain as objective as possible; yet one
> must give approval to that which seems worthy of approval.

But a few months after that, he leaves Hamburg on a ship for New York. From there he takes the Greyhound bus back to Whittier. Now he has a new plan—to go back to Berkeley in the fall, switching his major from English to sociology, a move he will explain by saying that Shelley and Byron hadn't prepared him to argue politics with the young Germans at the *komaradschafthaus*.

Along with his memories and doubts, he carries in his luggage a dagger inscribed with the Hitler Youth motto, *Blut und Ehre*—Blood and Honor. Later this souvenir will disrupt his life and come very close to destroying his future.

7

THE MUTE YEARS

With his return to the United States, just in time to draw the curtain on another series of terrible emotional blows, Jack broke off the stream of letters to George Chisler. I think he must have spent the summer of 1937 at home with the stepfather he despised. His brother Leonard would have been there too, another big young man rattling around that small mustard-colored house. Knowing his love of fishing, I'm certain he tried to escape by hiking into the Sierra Nevadas with his uncle and cousin Ray—perhaps the trip Ray was remembering when he told me, years later, about the time Jack came back from Europe so determined to fish, he hiked up into the mountains in a pair of black dress shoes. "I still have an image of him," Ray laughed. "He had waded out into the frozen water, no waders on, turning blue, broken glasses, flailing away."

By September he was back at Berkeley, living in a single room with a bath down the hall and working on a master's degree under the supervision of a tyrannical old sociologist named Frederick J. Teggart. He described this part in one of my college letters, how he plunged into history and political theory and economics and statistics and anthropology, weighing Teggart's preference for pure social science against the rival Chicago school of "muckraking" sociology and finding himself more and more skeptical of intellectual fashions and trends. At the end of his first year, Teggart tapped him as the department's first and only paid teaching assistant, a considerable honor.

Then this "happy, cloudless time" came to an abrupt halt:

> I received a telephone call one afternoon from home announcing that my brother had been discovered dead of a gunshot wound in the basement of our home. He had been cleaning a rifle. He had shot himself through the heart. There was no note nor any indication of any kind of intentional death.

Too distraught to face anyone, he left a brief note at the sociology department and rushed home. What happened when he got there, what his mother did and what role his stepfather played, I cannot say. At this point the letter tamps down into the disciplined style of his best intelligence reports:

> My brother, Leonard Albert, was about 6'2" and good-looking. He went to Whittier High School and Whittier College, about an average student, no special talents, no vices, had difficulty finding himself. He didn't have a job, didn't know whether he wanted to stay in college, and hadn't found his way with the opposite sex. He was a decent human being and I loved him very much.

It seems significant that in the very next line, interrupting his narrative in such a removed and uninflected way that on first reading I barely noticed it, he mentions his stepfather for the first time—"a car salesman named William Beech" who had been living hand to mouth and was "looking for an easy berth." He had no respect or liking for this man, and neither did his brother.

Then he turns back to the aftermath of Leonard's death. His mother bore herself with her "characteristic stoic courage," he recalls, drawing a lesson that would color the rest of his life:

> There is no answer to accident and death except courage
> and endurance—no other way to extract even a small dignity
> from the inevitable defeat.

Then an odd thing happened. A few weeks later, as he arrived back at Berkeley, before he had a chance to stop in at the sociology office to report in, he happened to see Teggart on the street. Teggart "cut him dead." When Jack spoke to him, he ignored him.

What did it mean? Was Teggart angry because he didn't ask permission to leave? Because he'd missed so many classes? Because he didn't go to him for solace?

There was no way to tell. But it was upsetting, very upsetting. Teggart wasn't just his mentor but also the head and founder of the sociology department. His whole career was vested in this one inflexible man.

Drawing on his mother's example and his growing reserves of stoicism and self-control, Jack stifled his feelings and finished out the semester. Then he went home to plan his next move.

Before he got a chance, his mother asked him to take her to the doctor. This was odd. As a nurse, she knew her way around the hospital much better than he did. That's how he happened to be sitting by her side when the doctor told her she had breast cancer.

> As a registered nurse, she knew the verdict was death, and
> she knew the detail of the course this malignant illness would
> take. Her reaction was—no tears, no protestations, no weep-
> ings and wailings.

When they got home, Jack slumped to the living room floor and wept for both of them. But his mother went on as usual, playing bridge and going to the movies and visiting her doctor for a weekly "X-ray treatment." He spent the year tending her at home, commuting to USC to pick up a teaching certificate and prepare himself for some kind of professional life. At one point, they gave him a psychological test. "I came out 87% introverted, which the tester said was "clinically interesting."

And here my college letters trail off. Maybe I hadn't responded, maybe he lost the thread, maybe the next part of the story was too muddled or too painful. I know that he got his teaching degree and taught for a year at a small prep school in Santa Barbara, where he offended one parent by giving his teenagers advanced books like *Brave New World* and became the target of a brief loyalty investigation after his landlady saw his Hitler Youth dagger in his room—a pair of FBI agents came to interview him and quickly decided he was no German spy. When he could, he went home to check on his mother. Then the Japanese struck Pearl Harbor and he drove straight to the enlistment office, only to be rejected for bad eyesight.

By this time, he was passionately anti-Nazi. He tried the merchant marine. They wouldn't have him. He tried the FBI. They didn't want him either. Finally he decided that if he couldn't join the fight, he had to do something bigger than teaching in a prep school. So he enrolled in the University of Chicago, aiming for a Ph.D. in sociology.

Of his state of mind there, I have only one solid piece of evidence, a letter from a young anthropology student named Ray Birdwhistell. Written a year later, it begins with warm memories of the days they wandered around Chicago, talking philosophy and visiting the Negro bars on the South Side. But the bulk of the letter imagines a typical conversation they apparently had many times. "Visualize us talking," Birdwhistell orders. "You on the hard red couch and me in the blue $3.78 armchair, drinks before each of us and two rapidly disappearing packs of cigarettes."

> JACK: I am a mute.
> BIRDWHISTELL: The hell you are! What you call muteness could better be described as lacking the touch of the dramatic. You are capable of far more than anyone I have ever met.
> JACK: But Ray, you can write—you can feel—you can express. None of these can I do.
> BIRDWHISTELL: Jack, fa' crissake! You must raise your evaluation of yourself. You shall shine! You shall shine!"

On it goes, page after page, slopping over with affection. Someday the Hopi and the Zuni will never again be mentioned in the same breath, Birdwhistell swears, all because Jack teased him so many times about contemplating his navel. "God, what a lovely catalytic agent you are! My identification with you is so strong! Let's have another drink!"

And with that Ray Birdwhistell disappears, off to begin a distinguished career with what did, in fact, turn out to be groundbreaking work on the

Hopi and Zuni. Later he became known for his long association with Margaret Mead. But Jack saved no other letters and never mentioned him.

Again, life interrupted. First the military changed its rules on eyesight, reclassifying him 1A and giving him ten days to report to the Whittier draft board.

When he got home and saw his mother, he got another shock. Later I found a terse description in a letter he wrote to my sister: "The cancer had spread throughout her body and had entered her skull."

She was dying. He told the draft board she didn't have long, but they ordered him to report for duty anyway.

Then he went home and a few days later their doctor came to the house, walking heavily up the stairs. He said later that he believed the doctor gave her something and had always been grateful for this act of mercy. For my sister, he summed the terrible moment up in a single line:

> I awoke in the morning to find that she had died in her
> sleep, jaw hanging slack.

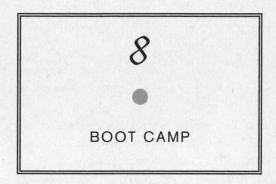

8

BOOT CAMP

Here the letters to Chisler resume with a vengeance, every week and sometimes every day. After burying his mother, Jack steps onto a military bus and soon arrives, on the afternoon of January 28, 1943, at the brand-new Arlington Reception Center just outside of Los Angeles. Before the day is done they've given him an IQ test and tests for mechanical aptitude, plus a raincoat because it's raining like hell. Then on to language exams in French and German and "an excellent film on venereal disease" plus a little marching and barracks scrubbing, activities he finds surprisingly pleasing. Of course, he continues, there's no privacy. But that's to be expected. And the food is good and the officers and sergeants treat him well. There's even a little theater that shows brand-new hits like *Casablanca*.

Five days later, he leaves on a troop train for points unknown. And what a pleasure it is, seeing everyone give a hundred percent to the war effort, the USO women who come to the train stations to sell them candy and Coca-Cola at cost and the farmers who can't wait to stop their cars to give a soldier a ride. But the real revelation is the life of the recruits:

> The main activities of our trip have consisted of poker, blackjack, crap shooting, profanity—every other word is a four-lettered one—and looking at every girl that passes with remarks on what hot or cold stuff she is.

Seven days later he arrives in Florida and discovers he's been assigned to the Signal Corps. Billeted in the Sherman Hotel in St. Petersburg, he spends the first day cleaning his room over and over until the sergeant is finally satisfied, then starts memorizing the eleven general orders. The next morning he gets up at five and drills until sundown. The next day it's the same. It's all about climbing poles and rigging wires, muscle work for people who couldn't pass the Radio Aptitude Test, so it's no particular distinction to have been chosen. Plus they have the highest mortality rate in the service since the enemy is always trying to cut the communication lines.

But again, there are bright spots. There's the sunshine, the ocean, dances with local women, a few more useful lectures about venereal disease and an excellent movie about the differences between democracy and Fascism. Plus his very own gas mask.

Two weeks later, the Army moves his unit to a camp on the outskirts of St. Petersburg. Now they're going through basic training, up at 5:30 for roll call and breakfast, then an hour and a half of calisthenics, then drilling and rifle-training and lectures on first-aid, military courtesy, morals and poison gas. They're living out of duffel bags in eight-man tents with sand floors, and the surprise is how cold it gets here in north Florida; that first night he wore his winter and summer underwear and fatigues and flight jacket and overcoat too. And the griping is getting on his nerves. The soldiers gripe about eating out of mess kits, about digging through their duffel bags, about the sand floors and the bugs and every other damn thing. "I don't know what they expected of army life—apparently they want the comforts of home."

But as the weeks go on, he starts to gripe a bit himself, what with this bad cold and hellish sore throat and frankly, he's starting to get just a little bit tired of being surrounded by other people and their germs all the time. And the food isn't so good and when it comes to floors, George, sand is *unbelievably* worse than grass. And the lines! Line up for meals, line up for inspections, line up for passes, line up and wait and then wait some more. He's sharing a tent with three Mexican boys named Rodriguez, fine fellows all, but there's another Mexican named Peña who is the most disliked man in the camp, a smart-alec tough-guy loud-mouth goldbricker who keeps complaining about discrimination—which is complete rot, George, typical of this interracial business. The fact is, the fellows dislike Peña because of his qualities as a man. "So far he hasn't bothered me but if he does, I'm going to fight him without ado even if I do get a good licking."

A few more weeks go by and he starts to adjust. They just had a hurricane warning that was fun, learning how to lie flat on the ground holding the corners of the tents. Then they went for a long hike in the rain and that was kind of fun too. He even enjoyed standing in a long soggy line to buy candy bars and milk. In fact, he's glad to be in Tent City with sand floors and so much "flora" everywhere instead of some requisitioned hotel room. This is the real Army life, sitting on a damp cot in wet shoes and wet socks and wet pants and a wet shirt, writing a letter by candlelight. And he's really liking the Rodriguez boys. At night in the dark tent, they sing Mexican songs very well indeed. One night, Paul Rodriguez admits he's hoping they won't have to go to the front. "Maybe we won't," Jack says, keeping his real feelings secret.

A few more weeks pass. By now, Jack has a flat belly, hard muscles, and a deep tan. He's excited about the possibility of a job in cryptography. But he's still puzzling over his fellow Americans, who seem to have such strong pride and loyalty to their individual states but so little for the country as a whole—no national feeling. And such disgusting habits, like the constant stream of obscenity that revolves relentlessly around defecation, urination, copulation, masturbation, and homosexuality.

> Having seen man in the mass, I find that I'm not favorably impressed with him. He seems obscene, discourteous, unrefined, and particularly unreflective. Jonathan Swift's picture is born out in every detail.

But maybe he's just in a sour mood. For the last month, he's been able to read only a single book—Harold Lamb's *Ghengis Khan*. He's also been picked up twice by the MPs for drunkenness, though he doesn't mention it for a few more letters. And there's something else troubling him, a crisis so disturbing he hasn't mentioned it to Chisler at all: Beech's lawsuit against his mother's estate. Apparently Beech is not only demanding money but also threatening to stir up some kind of dirt. The evidence is a letter Jack writes to his attorney:

> I appreciate the sober caution with which you approach the Beech affair, but because I hate the man I categorically refuse to show the worldly wisdom to deal with him. I'm willing to accept whatever consequences that may ensue from this decision but I will not change my mind.

Money is not a concern, he says. He suggests they use "the incriminating Beech letters," which summons an image of him receiving or uncovering the private correspondence of this hated man, a flaming moment of betrayal and discovery. He also mentions a few people who could testify in court, "*particularly the Whittier woman with whom Beech had an affair.*" But he gave no more details and never mentioned it again. These are the only existing hints of the hidden life of his family and the secrets that may have driven him from home, secrets that seem to have touched him in a place beyond reason:

> Please don't overlook anything here. I can't get a furlough to come home and deal with the situation. It might ruin every chance I have in the army.

At the end of six weeks as a soldier, Jack spends a rainy Sunday afternoon in his tent writing yet another letter. His Mexican friends have been passing the hours with hearty singing of Mexican songs and now three of them have lapsed (or relapsed) into siestas. Albert Peña is in their tent now—remember, George, the obnoxious loudmouth who kept complaining about discrimination? It turns out there are reasons for his aggressiveness. At eighteen his brother was sent to prison on a life term, and Albert suffered from that and also from "the social discrimination to which Mexicans are subject." But he's a friend now. He and the Rodriguez boys have even been kind enough to teach him a considerable amount of the foulest Mexican profanity, which he proves by listing seven curses in descending order from most offensive to least, adding male and female endings wherever appropriate. He doesn't even mind when Francesco and Pablo save themselves a walk in the rain by pissing in the tent. "I protest only good-humoredly, '*Pablo, por Díos!*'"

As the training eases, they go to the beach and laze around in the surprisingly warm water. The other soldiers are starting to ship out and with them go Pablo's wristwatch and his own Schaeffer fountain pen. But what really upsets a fellow is the complete lack of Army spirit.

> These are young men I'm living with, the so-called wave of
> the future. Yet I've never heard more than one or two ever state
> a political, social, economic, or ethical idea. They seem to have
> no particular interest in problems of Communism, Fascism, or
> democracy. There is no discussion of Germany or what we are
> fighting against. The men seem simply mute; they are for the
> most part culturally, politically, and sociologically illiterate. For
> a man who has an almost religious feeling for democracy, this
> experience of America's young men is discouraging.

Ten days later, Jack boards a troop train himself, six hundred men travelling west toward an unknown destination, rattling through Georgia and Tennessee, Arkansas and Kansas and Colorado, and on up through the beautiful Sierra Nevadas before pulling at last into a disbursement depot just outside of Fresno. Now this is a fine camp! A movie theater, a bowling alley, tennis courts, a decent barracks, and even a "really excellent library." And here Jack meets soldiers more to his taste, including a Harvard sociology grad who speaks four languages and a good-hearted young Romantic who taught literature at the University of North Dakota. But soon he becomes frustrated again. It's not so much that he's living in a tent with a slit trench for latrines, or the lack of hot water, or getting quarantined because his bunkmate came down with scarlet fever. It's not even the endless waiting.

No, the real problem is this damn Beech situation. Writing to Chisler at the end of April, he confesses how extremely anxious he is to get this job in cryptography, dropping another hint of his secret troubles:

> Because the position demands a sound constitution and men-
> tal stability, Beech's suit vs. mother's will on grounds that she was
> insane when she made it may have already ruined my chances.

On grounds she was insane? Is this merely Jack's alarmed interpretation of the "not in full possession of her faculties" clause, a gambit lawyers often use when contesting a will? Or is it another key to the mysteries of his family? He doesn't elaborate and never mentions it again. But it seems he has a reason to worry after all—already the Army has written to at least one of the people he listed as references, and sent an investigator to conduct a thorough interview with Ray Birdwhistell. If only George could give him some sort of reassurance!

> I'm anxious to know whether the army has written to you
> or interviewed you with reference to some such matter as this.
> Please let me know as soon as you can.

Although he doesn't mention it, he's probably also worried about the FBI investigation into the Blut und Ehre dagger. He's relieved when George writes back and says he gave the investigator a glowing recommendation.

Jack spends the summer marking time, getting training and more training. Sent to a small base in Porterville, he learns to climb poles and splice cables and keep a field switchboard. He learns to sit in a room filled with tear gas and take off his mask, letting the gas sting his eyes while walking slowly around the room. Then he practices putting on the mask. He takes a familiarizing sniff of the four deadly gases most often used in battle: chloropicrin, mustard, lewisite, and phosgene. For diversion there's dancing at the Elks Club and all the interesting Americans who pick him up hitchhiking.

From Porterville he moves to Visalia, where he makes friends with an aspiring young novelist named Dick Hagopian. In a beautiful camp built around a running stream, he learns codes and secret message transmission and map reading and resolves to spend every night studying in his tent, "except Saturday evening during which I intend to frequent a dance or a worthwhile show."

He's happy with everything now, he tells Chisler. The macaroni and cheese for lunch, the steak for dinner, the bottomless coffee and jam, the foothills rich in oranges and olives. Why, the oranges in Visalia are the best in the entire state. And yesterday he put a dollar into a crap game and came out with twenty-three. He encloses pictures of himself with a few healthy Visalia girls and also a woman who picked him up hitchhiking because her son was already out in Europe fighting so she never passed a soldier without picking him up. So who cares if the life expectancy of a radio operator tends to be extremely short? It's an important job and he's proud to do it.

And then it happens, the moment that changes everything. His sergeant tells him to pack up and ship out, and for the first time he's travelling by himself. They won't tell him where but he's headed in the direction of Bakersfield, he tells Chisler in a dashed-off note. Maybe this is his big break. Maybe the cryptography job is finally coming through.

But it isn't a job in cryptography.

Two days later, he sends Chisler a mysterious postcard from the Glacier Point Hotel in Yosemite National Park:

> Everything turned out exactly as I told you it might during our ride. Am now by myself at a rail station to make a long trip. You may not hear from me for some time. But know that everything is *exactly* as I told you.

Then he rides the train down through the Sierra Nevadas to Atlanta, where the Army issues him two suits and orders him not to wear a uniform—and to keep his mouth shut.

> I've been told that I can inform trusted friends that I am in the Military Intelligence Division. However, they are not supposed to know what branch I'm in nor are they supposed to tell other persons what I'm doing.

Soon, probably within a few days, he'd probably be sent somewhere nearby for field work and then more school. With any luck, he'd be ready to go with the troops when they invade the continent.

He stops there, obeying his order of silence for the first time.

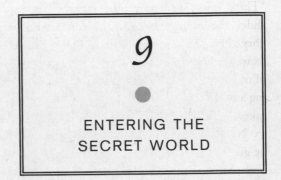

9

ENTERING THE SECRET WORLD

This is what happened, as best I can tell. Somewhere in some secret office with an unlisted phone, a military officer who was probably not wearing a military uniform sat my father down and asked him a fateful question. "How would you like to roam battlefields at night picking documents off of dead bodies?"

There would be no promotions, the officer continued. The work would be dangerous and secret. It would require enormous discipline and loyalty.

He described a group called the Counter Intelligence Corps, one of the most secret of America's secret agencies. There had been quite a fight in Washington over whether it should even exist. It took more than two years after Hilter invaded Poland to overcome resistance. Later, General Dwight Eisenhower would call it a "shocking deficiency" that was allowed to persist because "the American public has always viewed with repugnance everything that smacks of the spy." Only after Japan bombed Pearl Harbor did the War Department finally shake out funds for the first five hundred agents, who faced daunting requirements designed to keep levels of professionalism as high as possible. The minimum acceptible IQ was 120. They had to have special language or technical skills. Nine out of every ten applicants were rejected. Promotions were banned because giving such men a fair shot at climbing up the ranks would mean losing too many to the officer corps.

My father said yes, joining an extraordinary group of recruits that included men like Henry Kissinger and J. D. Salinger. Many years later, a retired CIC agent named Arthur S. Hurlburt described a typical first day: "There was a lieutenant by the name of Cameron who showed movies and slides of bombings and knifings and other terrorist activities with all the gory details. When the films were over he came out on the stage and told us we were starting in a career of just what we had been looking at. If any of us had weak stomachs, now was the time to find out. The bus was waiting and would take anyone back where they came from, with no questions asked. Sure enough, a handful of people got up and left."

Not my father. Picking up the story in a new batch of letters to Chisler, he gives up drinking and dancing and plunges into his training: rapid fire, slow fire, bobbing target, shooting from the hip with his .38 revolver. Then the .45 and the carbine and the shotgun and the .78 and judo instruction. Practicing his new professional secrecy, he doesn't mention the required courses that other CIC men described after the war: lock-picking, surveillance, planting bugs and phone taps, photography, interrogation and undercover work and the use of invisible ink. They're taught to always leave something in a room placed in a way they can remember—a dime between two handkerchiefs in a drawer, a hair on top of a light switch, so they can tell if someone searched their room. They attend lectures on police systems, frontier control, and existing CIC operations. Repeatedly, their instructors tell them they're going up against the best—the Japanese Kempei Tai has ten years of experience in war, the Gestapo four years, and the Russians *two hundred years* of organized spying in war and peace. So they have to be better.

It's daunting, and sometimes Jack feels a little dubious about measuring up to such expectations. He tells Chisler it might be a good thing Atlanta has few pleasures to offer beyond its many beautiful women, "a pleasure which I cannot now take indiscriminately." But at least he has time to read, to bolster his spirits with heroes from the American past. Right now he's in the middle of Padover's *Jefferson*. "To me, Jefferson is still the great spiritual leader of this country and constitutes a set of values which can make life worth living to a man."

There's also political training, and already this intelligence service is reinforcing his steady march from collegiate socialism to anti-communist. His study materials include a global threat analysis from military intelligence assessing the relative dangers of socialism, Fascism and Communism. At a time when Hitler's troops are advancing on Moscow and Stalin is eagerly cooperating with the Allies, the CIC's specialists are reminding their recruits about times like 1919, when the Soviets encouraged American troops on the Rhine to mutiny and Lenin created the Communist International, of the show trials of the 1930s and the infamous nonaggression pact with Stalin that gave Hitler breathing room to start the war. Soon CIC agents become so passionate about their "countersubversive" work that one of them will try to bug Eleanor Roosevelt's hotel room in hopes of catching her in bed with a military officer suspected of Soviet sympathies. When President Roosevelt retaliates by shutting down the CIC's countersubversive program, eight out of the nine CIC area commanders will violate the president's order by hiding their files instead of destroying them.

My father is definitely catching the fire. Writing to Chisler a few months into his training, he describes a disturbing incident at an Atlanta hot dog stand. As he waits for his lunch, a small Negro boy approaches the counter and the white counterman snaps at him, ordering him away from the counter. "Conditioning for the negro begins harshly at a very young age," he muses. Then he turns to the geopolitical implications:

> If the Communists ever become a strong group in this country, consequent upon a possible Russian expansion in Europe, Africa, and Asia, I think the negroes will go over to them.

Chisler lets this pass and encourages his racial sympathies with a story about life back home in Los Angeles, where sailors and soldiers have been attacking gangs of Mexican "Zoot Suiters." Instead of punishing the sailors and soldiers, the city banned Zoot Suits and arrested the Mexicans. A more radical response comes from George Ulner, one of the Whittier College friends who accused him of intellectual dishonesty years before. Now under the influence of a British Marxist named R. J. Lasky, Ulner writes to express his surprise at Jack's negative remarks about communism. "You know how important Russia is to the repressed worker. What people really want is freedom from the fear which is the direct corollary of a system of free enterprise, labor suppression, race prejudice and imperialism. We must not listen to the sour and prejudiced voices that would prepare us for the greatest of wars with the very nation that might have saved our national integrity—Russia."

That's the end of his correspondence with George Ulner.

Late that June, Jack finally gets his orders. Writing to Chisler, he says he's going overseas to a destination he can't reveal. They might not see each other again for many years.

> I suppose that this is the time to speak frankly. The loss of my mother and brother continues to hit me pretty deep, and I think this accounts for my restless desire to keep moving. Also, I feel a loneliness because of having no wife or family of my own. These things contribute to my not particularly desiring to avoid danger. In certain sad moments, I feel that I should prefer not to return from this war. Yet I recognize this feeling as a completely unworthy weakness. The battle of the post-war world will be almost, if not equally, as important as those of the war itself, and I believe it is the moral obligation of each of us not to step out of this struggle for a better world.

A month later, he posts his last letter from the United States. Increasingly secretive about his real work, he says he spent the last few weeks going through "an intensive course of instruction about which I can tell you nothing." There isn't much else to say. After an eighteen-mile hike with a full field pack, he feels marvelous. The picture of baby John Chisler in his billfold—he'll carry it with him when he goes across. In the event he doesn't return, he wants George to have his books. The rest of the letter is about his tour of Washington, D.C., where he visited all the monuments and even the tombs of George and Martha Washington.

> I made a particular point of visiting Jefferson's monument—very beautiful—with a great statue of Jefferson standing. There I re-read the Declaration of Independence and the

statement inscribed on the memorial: "I swear upon the altar of God eternal hostility against every form of tyranny over the mind of man."

No wonder Chisler shoots back a reply fretting about the dangers of idealism. "Remember—don't try to win the war the first day."

10

THE WAR BEGINS

One day many years later, I find myself in possession of a box of my father's most personal keepsakes, mostly pictures and a small cache of Victory Mail. Postcard-sized photostats of letters shot on microfilm on the front and shipped to developing stations back home, the V-mails are frozen pieces of the past. The first is dated September 27 and posted from a replacement camp in Canastel, North Africa:

> Am now living in a tent with other members of my organization on a fairly barren windswept hill harassed by sun and dust storms. I will be moving again very soon I hope into a combat unit.

They tried to post him to Tangiers, he continues, but he refused. Just two weeks before, troops landed on the beaches of Naples and a CIC unit hit the beaches with them, wading through bodies and wreckage under fire. *That's* where he wants to be.

His mood gets worse as the weeks drag on. He's filthy and hungry and sick of waiting. He talks to an experienced British soldier about what it's like to see battlefield gore and the soldier tells him that in England, training "includes a gradual toughening of soldiers to blood by cutting pigs throats." Another soldier starts to rub him the wrong way and they squabble so much their commanding officer finally orders them to take it outside. As the other members of the unit improvise a boxing ring and gather on the dusty hillside to watch, Jack fights with such dogged disregard for pain or injury they decide he's the classic bullheaded Scotsman and christen him with a classic Scottish nickname—Jocko.

And finally he gets his wish. When I convince him to resume his series of autobiographical letters, this is where he starts, grumbling a little. "I approach this exercise with some skepticism as it seems to me that my past experiences are probably of little interest or meaning." Then he remembers the sweeping curve of the bay of Naples and all the colorful buildings gathered at the shoreline, the great adventure beginning at last. Things are moving so quickly, the unit he's been sent to join has already moved ahead to Caserta just thirty miles south of the German lines. As he races to catch up, he can hear artillery in the distance. He asks where he can find Unit 305 and gets pointed up into the hills above Caserta to the Palazzo Leonetti, a sprawling but filthy mansion nestled on a beautiful hillside. He opens the door and stands there for a moment, taking in the huge empty room. Then a sergeant lounging next to a pot-bellied stove calls out in a commanding voice, "Corporal, shut that door!"

What an arrogant SOB, he thinks.

Next he meets his new commanding officer, Major Stephen J. Spingarn. A stout and eccentric man who worked as a lawyer for the Treasury Department before the war, Spingarn is in a fury that day—I got this part from the CIC records in the National Archives—because they discovered that an Italian man carrying a signed and stamped pass from another Army unit was actually an enemy spy. What idiot failed to observe *the simplest elements of security?*

That moment sets the themes of the next two years: vigilance, caution, security. Immediately Spingarn makes these themes concrete by sending Jocko out to look for some Fascists who are supposed to be holed up in an ancient mountain village named Caserta Vecchia. When that comes up empty, Spingarn orders him to investigate a soldier who's been sending letters to a German town, a possible sign of treason and espionage. That turns out to be another bad lead. Soon the records of the 305th CIC show them settling into a routine: "Smith, Richardson and Kern reported on a security check of certain installations . . . Smith and Richardson reported that they had turned a private and a merchant seaman over to the MPs in Naples. The two men had been quarrelling violently."

But arresting drunk soldiers and checking fences isn't very fulfilling, and it doesn't take Jocko long to see that he's not going to get very far asking after Fascists in primitive Italian. So he studies his intimidating new commander. Like him, Spingarn has a scholarly background—his father was a distinguished literary critic at Columbia, founder of The Spingarn Memorial Medal for Outstanding Negro Achievement—and already he's built on that background by making connections with prominent Italians like Count Carlo Sforza and the writer and politician Benedetto Croce, people who can act as guides to the local issues and personalities. So Jocko develops his own local source, striking up a conversation in French with a liberal Italian lawyer named Antonio de Franchis. Downing glass after glass of wine, they talk about classical literature and philosophy and eventually de Franchis confesses that his professional career has taken quite a lot of punishment due to his weakness for feminine beauty. Not only does he have a number of mistresses and illegitimate children, but the husband of one of his mistresses has just been placed in an internment camp. He's working every angle he can to get the poor man out.

But de Franchis knows everyone in Caserta, and now Jocko starts making some real progress. He arrests a dangerous Fascist named Stefano De Simone. He develops a dossier on a politician who might be financing a neo-Fascist political movement. Relaxing into his work, he begins to make friends with his fellow CIC agents. One of his favorites is Gordon Messing, a rabbi's son with a Ph.D. from Harvard in classical philology. Short and unassuming but with quick eyes that hint at a powerful intelligence, Messing also has such a quileless personality that the others wonder if he's really suited for intelligence work. At the moment, he's carrying around a Sanskrit-Flemish dictionary.

Another favorite is Gordon Mason, the same obnoxious sergeant who barked at him to shut the door. A former newspaper reporter from Ohio,

Mason can build furniture and repair cars and whip up a salad out of dandelions he picks in the field. He's handsome, debonair, sardonic, charming. Women love him. He's the one who organized the domestic staff of the Palazzo, six or seven Italian POWs led by a former Fascist. "We will treat you well," he told them. "You will eat what we eat. You will live with us. But if you ever serve us Spam, I will have you all shot."

Jocko and the two Gordons become best friends "in war and against the world," as he puts it later. All are productive agents whose names appear frequently in the files. The two Gordons also have the comical job of investigating the mistresses of leading Allied generals, along with various prostitutes and the more suspicious nuns. The CIC records detail one report they did on a woman named Ada Simis who followed a German soldier to Caserta but switched her interest to the new American troops. "It was the opinion of the agents that her position and capabilities are not such as to render her a menace to Allied Force security. She is evidently more concerned with sexual intercourse than espionage."

By late October, Jocko's letters to Chisler have a new confidence and energy. He describes passing through bombed cities and drinking delicious wines and Italians who are very friendly to their faces but always deny having anything to do with Fascism. What he needs now is a good English-Italian grammar book. "I'm having the greatest time of my life. It will be better when I get a chance to fight with a gun."

A month later, the Germans have withdrawn to their winter positions along the Gustav Line. Rain and snow are coming down constantly. Now he's in the thick of things, talking to everyone and learning first hand a million fascinating things, like how hard it is to keep a village running when bridges and railroads and buildings everywhere have been bombed into fragments and stores have been ransacked and there isn't much food and the little bread they find tastes worse than sawdust and all the young men dragged off by Germans. He agrees with his "liberal friend of wide views" that there seems to be little hope for Italy. There are no leaders on the horizon to lift it up. Everyone is venal and corrupt, yet the people are so warm and friendly. He has discovered that he feels no hatred for the enemy, none of the bitterness of war. More than ever, he believes in education and in service. At the same time, he knows that the Fascists and collaborators have to be punished. "Please, George, don't think I'm going soft on collaboration or compromise."

I'm not so sure that the war will be over soon, but I feel now as I always did, glad to be in and willing wholeheartedly to stick it out to the end no matter how long it lasts.

In his letters to me, years later, my father reduced all this to tidy lessons. Meeting de Franchis taught him the importance of developing deep relationships, he said. "It would be hard to exaggerate the value of a really key, central contact in any field." Working in the villages made him wonder if the Italians would ever see that their rampant factionalism "was what sparked the chaos that led to the rise of Fascism in the first place—anarchy leading to tyranny." That's Aristotle's Iron Law of Politics, son—a society needs a minimal foundation of order and if it isn't achieved by cooperation, it will be achieved by coercion.

Even his love life became an opportunity for instruction.

> In Caserta I also fell in love for the first time (age about 30) with a young Italian woman from Milan. She was a baroness, married, two young children. Her husband was an Italian Fascist officer and a prisoner of war in North Africa. Thinking of her and of my years in Italy, I can say that Italy and Italian women contributed considerably to whatever civilization I've acquired. For more on this theme see George Santayana's one and only novel entitled *The Last Puritan*.

He didn't mention the time he reported a buried cache of German weapons or finding the location of a German minefield on the outskirts of Rome—I found those details in CIC records. And when it came to the physical dangers, he insisted there were none. He remembered just one scene of violence, when his friend Tom Lucid went to arrest a notorious war criminal at home. With his hand on his Beretta and a GI escort at his side, Lucid knocked on the front door. After a moment's silence, the door swung open and the man came out with a gun in his hand. He shot and missed. Lucid aimed his Beretta and pulled the trigger but the Beretta jammed.

Just when things looked bad, the GI raised his Tommy Gun and opened fire, saving them the trouble of a trial. But of course, my father insisted, he wasn't there for any of this. "I remind you again that I did not choose my kind of war; it was thrust upon me."

Later I learn that his unit lost seven men in Italy. The first was an accident—John Walcott of Boston, killed in Caserta by a passing truck—but two men were shot at Monte Cassino and another at Anzio and two more in the last push at the end of the war and one more after the war ended. I get another glimpse of the grimmer side of things from a series of articles Spingarn wrote for the *Saturday Evening Post* after the war. Sailing to North Africa as an easygoing American idealist, he landed in a snakepit of social intrigue with Spanish Fascists and Vichy French who pretended to be allies while concealing secret ties to the Nazis. Then he led his unit ashore under fire and discovered that most of their training was hopelessly naïve. All that stuff they dragged ashore, the handcuffs and wire-tapping kits and blackjacks and cartons of moulage for taking tire and foot prints, it was all useless. Pushing up to Naples, fanning across the ruined city securing documents and arresting Fascists, they discovered that what they really needed were Jeeps and Italian interpreters. Before long Spingarn was pushing himself so hard that once he drove his Jeep right through enemy lines into the fire of a machine gun—so hard that later his men would come close to mutiny.

They were on their own, figuring it out for themselves. People were dying just up the road.

11

CATCHING SPIES

After six months in Italy, Jocko posts another V-mail back home. Now it's March of 1944 and he's back in the Palazzo Leonetti after a few months of travel, writing before an open fire, eating excellent food prepared by Italian cooks—turkey, sugar-baked ham, toast for breakfast, plenty of butter, bread, marmalade and coffee. At the moment, he is listening to the music of the Italian radio. He has almost complete freedom, a Jeep to travel in, has been hunting twice with shotguns for duck and plover, and is thinking about a trout fishing trip in the near future. He describes Mason and Messing and the other CIC men, journalists and lawyers and even a Rhodes scholar, evenings spent playing poker or shooting craps. Out of shame or respect for military secrecy he gives no hint of the battle still raging just thirty miles away at Monte Cassino, where the Allies have spent the last two months trying to break through the enemy lines at the cost of fifty-four thousand men—close to the number of Americans lost in the entire Vietnam War. At Anzio the big German guns and German planes are still playing turkey shoot with the trapped Allied soldiers, killing a total of seven thousand over four months. But he does say that he just got back from seeing a Franchot Tone and Deanna Durbin movie called *His Butler's Sister*, and he has such good Italian friends and "one good girl friend" and *none of this is his fault.*

> For the record and so that you may understand my position I have the right to say that about three weeks ago I asked our commanding officer to transfer me to a front division; I asked him earnestly but he gave me no encouragement. When the European war is over I promise you and my own conscience to volunteer insistently for service in the Oriental war. I've done my best in these matters and the comforts I'm perforce surrounded by are no fault of mine.

Then things change. Setting out in rowboats from Gaeta, up the coast in German-controlled territory, desperate refugees start streaming across the battle lines. Landing on beaches, they flee into the towns and countryside. And with the refugees come spies. In a town called Mondragone, an Italian guard stops a man and finds a radio hidden in his bag. A CIC unit picks him up and puts him through a fierce interrogation. Finally the man breaks down and admits he's a spy for the Germans. With a little more pressure, he admits that he wasn't alone—six other men came with him. So the CIC agents fan out and pick up the other spies, confiscating two more radios along with codes and frequencies and fifteen thousand dollars in Italian *lira*. After putting them through detailed interrogations, the CIC agents begin piecing together a picture of their enemy:

In Rome, the Nazis have stationed two units of the German military intelligence service called the Abwehr, one specializing in espionage and the other in sabotage. The German topography school is at the Albergo Dolomite, the espionage and sabotage school in Corredo. Soon they know almost every Abwehr staff member by name. They even learn their favorite bordellos. Now there's an advantage to having scholarly men like Messing on staff. Though he's far too mild-mannered even to think of using violence or torture, he turns out to be an unusually persistent interrogator, with a talent for noticing tiny inconsistencies and stray details. Fascinated by an odd twist of Italian grammar in one interrogation, he makes the suspect repeat himself so many times the bewildered man finally blurts out the truth.

Even the competition is impressed. In the wake of the wave of arrests and intelligence reports, the top British counterintelligence man in Italy issues a statement saying that the CIC has "proved itself beyond a shadow of a doubt to be a major intelligence organization."

But as spring arrives, the Allied troops break through the Anzio lines and join the forces marching on Rome. Jocko dashes off a quick note to say he's "in the field" living out of his Jeep and a sleeping bag. As usual, he's in little danger beyond the slight possibility of running into a mine. And while the boys at the front have to make do with bad cognac and bad wine at exorbitant prices, he's come into a bottle of Canadian Rye, some English beer and fifty cigarettes.

Years later, writing to me, he will remember passing through the smoking battlefields, how oddly gorgeous and peaceful Rome was on the day it fell, all the young men killed in the bloody battles he missed. But now he's too busy to linger. Given a list of enemy agents to arrest, he rushes off to pick up a Hungarian diplomat attached to the Vatican.

> With graceful courtesy he invited me into his living room to meet his wife and have a sherry. We fell into a fairly long conversation and I delivered myself of an unnecessary discourse on Jeffersonian democracy. I then took my prisoner off to headquarters—no weeping and wailing on anyone's part.

Within two days, they arrest so many spies and collaborators, they have to hold them in a park because there's no room in the jails. They find a German torture chamber filled with finger screws and meat hooks. Under interrogation, one German spy reveals that he was told to stand on a corner tossing a coin as a signal. So Spingarn sends a ringer to toss a coin and they pick up another eight or ten spies.

Then the Fifth Army starts pushing north fast and hard, and my father gets orders to stay behind in Rome and recover documents. Two weeks later he's on the move again, first to Acquapendente to do a political analysis of about twenty villages and arrest more local Fascists, then on to Grossetto and Rosignano Solvay and Volterra. By now they've honed their methods. Two teams of agents approach a town from different angles, driving slowly to keep the dust from the dirt roads from alerting any Germans who might be watching. Then one team hits the town hall and tries to make contact with local partisan leaders while the other team searches for enemy command posts and shuts down the local banks, telephones and telegraphs and mail services. With any luck, they end the first day with a provisional government in place and start interrogating collaborators on the second day. Working from town to town, they end up in the village of Tavernelle Val di Pesa just south of Florence, where Gordon Mason sets them up in a beautiful villa in Chianti country with acres of olive groves, a wine cellar and a walled garden filled with baby artichokes. They eat lunch on a terrace overlooking the beautiful hills.

But sometimes there's trouble, especially when villagers try to take revenge on local Fascists. If that gets out of hand things get bloody, so establishing the rule of law is the important thing. One day Jocko gets a report that the inhabitants of a nearby village have rounded up all the local Fascists and crammed them into the little local jail. Apparently the retreating German troops took the able-bodied men with them as cheap labor, and now a rumor has reached the town that the men resisted and the Germans lined them up and shot every tenth man. Tempers are ugly, violence imminent. So he sets off alone in his Jeep, driving fast through twenty miles of magnificent mountains and valleys. Parking opposite the church, he asks the villagers to assemble in the square. Then he climbs on the hood of his Jeep and addresses them in Italian, telling them he understands their situation but reminding them of Aristotle's iron law of politics, that anarchy and the lawlessness of violence lead to tyranny. "So I will take the three worst Fascists with me and intern them," he promises. "The others I want released from jail two by two and allowed to return to their homes without a stone being thrown."

And that's how it happens, the crowd of villagers moving aside just enough to let the Fascist prisoners thread their way to safety. And the three worst climb into his Jeep, greatly relieved.

By then it is August, 1944. Up north, the Soviet Army is pushing the Germans back across Eastern Europe. But as they approach Poland, the Polish resistance rises up and begins to attack the hated Nazi occupiers. They haven't forgotten what happened five years ago, when Stalin and Hitler signed their infamous "nonaggression pact" and divied up Poland, then rolled through Europe picking off fat little countries: Belgium and the Netherlands for Hitler, Finland and the Baltics for Stalin. And after Hitler made the fatal mistake of sending his armies east into Russia and Stalin switched sides, Hitler retaliated by leaking the story of the Katyn Massacre, when Stalin's secret police helped him slaughter some fifteen thousand Polish military officers in a bid to weaken the Poles and make them easier to dominate. So the Polish Home Army rises up not just to get revenge against the Nazis, but also to establish some kind of claim to Poland—hopefully the Allies will recognize their suffering and keep Stalin from crushing them all over again. The fighting is particularly fierce in Warsaw, where the Nazis unleash brigades of ex-convicts and SS police units who slaughter forty thousand civilians in a single day. And Stalin stops his troops at the Vistula River. He refuses to let the Americans and British use his airfields to drop ammunition or supplies. For sixty-three days, as the Nazis kill two hundred thousand Poles, the Russian troops sit in their tents and watch the slaughter.

To men like my father, hearing this news on smoky battlefields, the Soviets seem monstrous and the warnings of postwar conflict more and more prescient.

But he doesn't have much time to dwell on it. At a roadblock near Tavernelle Val di Pesa early that October, an MP searches a young Italian man and finds a large amount of fresh currency with the serial numbers in order, a sure sign of espionage. The MP calls the CIC and soon Special Agent John Richardson (a "sociologist from Southern California," in Spingarn's version of the story) arrives with agents Mason and Messing to interrogate the man. Where did you come from? How long did it take? Where did you stay? Name the towns. How many miles between them? How many farms did you pass? What kind of crops? The man gives them his name—Mario Martinelli—but insists he's simply an innocent bystander on his way south to buy vegetables. But when he tells them he lives with a countess, Richardson sends someone to crosscheck the story with the countess herself, then catches Martinelli in conflicting details. Finally he confesses, telling them he's a very bad spy but that his comrade would have much to tell them. "She is only a girl, but she has been kissed on the forehead by Mussolini himself."

As it happens, the CIC agents have already heard quite a bit about this girl spy. They even have a good description: "Age 18, height 5'1", thick set, black hair, dark complexion, Roman accent, claims to be skating champion.

Details of her mission are unknown except that she was to carry back copies of extreme left-wing newspapers." But over the next two days and nights of interrogation, Martinelli produces a sixty-three-page statement that includes the one clue that makes all the difference—she is wearing blue tennis shoes.

So they all hop in Jeeps and go out looking for her. Mason picks up the story here: near Pontepetri, so close to the enemy lines that there's not a person in sight, he spots a girl riding on the handle bars of an Italian boy's bicycle, her feet dangling. He catches a flash of blue. "What are you doing here," he asks her. "Do you realize you're in the front line?"

"I'm not in the front line," she answers. "I'm in German territory. I'm going to visit my family."

Glancing at the blue sneakers, Mason tells her he has to take her back to headquarters to check her papers. She rides in front with her hunting dog on her lap. When they get to headquarters and walk into the room where Martinelli is waiting, she doesn't show even the slightest flash of recognition. All that day Mason, Messing and Richardson take turns trying to trip her up, taking her through every detail of her trip over and over again. Then Spingarn takes over and does the same thing. They pass around a bottle of cognac to see if liquor will loosen her tongue, but she refuses to drink. They threaten to put out a story that she's a prostitute, that she confessed and turned in all her comrades. She tells them that their terrible behavior just proves what Mussolini always said, that the Americans are so cruel they would torment helpless young girls.

They do get one promising detail out of her—that she lives in Rome. Immediately they contact the Rome CIC and ask for a report. Over the next few days, they try to disorient her by leaving the light on all night, by giving her nothing but a single piece of toast to eat each day. Spingarn sits down in front of her and eats a huge plate of spaghetti. She glares at him but still won't say a word.

Then the local Italian counter-intelligence chief takes over, grilling her for six hours. That evening, Spingarn writes an interim report: "Various devices have been employed without success, including confronting her with Martinelli, who crossed the lines with her. I have virtually arrived at the conclusion that no short-term psychological treatment will persuade the girl to talk, which is a great pity since she is undoubtedly a gold mine of information."

Eventually they take a closer look at her personal effects, which include a white handkerchief. When they hold it over a candle words appear: ANHOERIGER DER ARMEE—LUFTFLOTTE ZU LEITEN. This means she's a member of the army and should be taken to the headquarters of the German Air Force. They confront her with it, and still she refuses to talk. Martinelli tells them they're going about it all wrong, they could get what they wanted much more easily if they just tortured her. Later my father tells this part to Chisler:

> He suggested we use hot irons applied to the nipples of her breasts and to the pubic area. He offered to carry out this torture himself. I smiled at him, shook his hand, and offered him cigarettes.

By this time, they despise Martinelli and admire Carla.

Finally, after five days and more than sixty hours of interrogation, they get a report from Rome CIC—twenty pages of detailed information on Carla and all the members of her espionage group. At last Carla gives in and tells them her remarkable story. Two days before Rome fell, she joined a group of female spies run by a German control agent known as Colonel David. After two months of training in Milan, she crossed through the Allied lines and collected intelligence on troop movements around Florence. And after a second successful mission to Rome, she was given a private audience with Mussolini himself, who praised her courage and presented her with an Iron Cross. Then she put on her blue tennis shoes and headed south on another mission.

A quick trial follows, with a British officer presiding. One thing that impresses them all, especially my father, is how thoroughly and even severely he questions them, almost as if they were in the dock themselves. He's especially interested in how they treated the two prisoners and what physical pressures they might have used. Although this puts them on the spot, it also makes them proud—it's justice, the thing they came to defend.

But in the end, the British judge finds both prisoners guilty. He sentences Martinelli to death. Because of her age and courage, he gives Carla twenty years.

The next day, Mason and Messing drive Martinelli to the firing squad. He's dressed in his best clothes. When they get a flat tire outside of Florence, he offers to change the tire if they'll take a letter to his mother. They accept the letter and let him change the tire.

Later Mason takes the letter to Martinelli's mother and she throws her arms around his neck, breaking into tears. Messing tries to discharge his feelings in his own way, fretting over Carla's psychology in his official report. "She is able to immune herself within the tight walls of her impervious fanaticism, caring nothing for other values and indeed despising them." Later he visits her in jail, still hoping he can make her see how wrong she is about Fascism and democracy. But she just laughs at him and says she doesn't understand why everyone is so interested in her.

"You're arguing like a little girl," he says.

"Well, I am a little girl."

Long after the war is over, Messing visits her one more time. By that time, she's the happy mother of a large family. She greets him warmly but still won't admit she was wrong. In fact, she's even become something of a national hero—in 1986 an Italian shipping company named a cruise ship after her.

She was a true believer, my father always says, sounding a little sad about it.

12

●

THE WAR ENDS

That November, Spingarn sends a secret wire to headquarters: "URGENT SECURITY SITUATION HERE. TEN SABOTAGE AGENTS CAPTURED WITHIN PAST FOUR DAYS THREE HUNDRED MORE REPORTED EN ROUTE OR TO COME SEVERAL CACHES OF ENEMY SABOTAGE EXPLOSIVES DISCOVERED . . . ADDITIONAL AGENTS URGENTLY NEEDED."

Fighting a last stand in the Apennine Mountains, holding the Allies back with thousands of interlocking machine gun nests and pillbox artillery units, the Germans managed to survive until the snow piled up and slowed the great Allied push to a crawl. Now they have begun sending a secret army of espionage and sabotage agents on a scale far beyond anything ever seen in any previous war. The German spies cross the lines wherever the firing stops or the troops thin out, travelling by foot or boat or parachute or simply staying put as the German troops retreat. They swim ashore in wetsuits, carrying bombs. They pretend to be priests or nuns, partisans or Polish deserters. One passes himself off as a doctor, another as a shepherd with an odd tendency to herd his sheep into tactical areas. So many walk right down the main road, the soldiers give it the nickname "Spy Highway."

Once again, my father moves into a nice private villa behind the lines. This time he has a cook, two Italian attachés and a civilian car. Writing to Chisler, he paints the scene with self-mocking precision, describing his easy chair and open fireplace, the warm wood fire and flask of red wine on the table. While men are getting killed in the snow and mud of the front, he walks around in civilian clothes and gets called "sir" by officers who outrank him. Every day, obsequious Italians inquire whether he's an *ufficial* or a mere *soldato*.

Did he mention that the villa was owned by a count?

But this time, though he gives no hint of it in his letter, he and his colleagues are working around the clock, making arrests and conducting interrogations and putting together their accumulating wisdom into an all-important document called the Pattern Report:

- German agents carry bills in serial order.
- They wear the same shoes and suits and carry the same suit-cases and briefcases. They carry identical sandwiches wrapped in identical paper.
- They sew their documents into their underwear or stitch them into their shoes.
- They tell similar cover stories, saying their houses have been bombed or they're afraid of being drafted or they're just out looking for their girlfriends.

Over the next seven months, they snowball these insights into an intelligence weapon so effective it helps them capture a total of five hundred enemy spies—more spies arrested in less time than in any other war in history. Most are tried quickly. About fifty are sent to the firing squad.

But as the winter drags on, Spingarn starts to unravel. He was eccentric to begin with, eating fourteen eggs in a sitting and bathing so rarely that the head of the Italian staff complained to Gordon Mason, "You got to do something about the Major, he never washes." Now he sometimes hides in his room for days, complaining of allergies and migraines. During morning staff meetings, he goes around the circle of officers citing each one of them in a hoarse whisper for various errors and crimes committed since joining the unit. Messing fell asleep on guard duty during the Atlantic crossing— that's a shooting offense! Barney brought his Italian girlfriend with him from staging point to staging point, against security and orders—send him up for court-martial! Later Mason will remember Spingarn showing up for the predawn planning meetings in a dress shirt and no pants, ranting with a clipboard.

Soon the soldiers begin to walk the olive groves of Tavernelle Val di Pesa, griping and scheming, close to revolt. Jocko urges caution. The way he sees it, Spingarn is suffering from isolation and overwork and also from his obsessive sense of responsibility. He feels a need to win the war single-handed. But isn't that something they can all understand? Isn't it an admirable flaw? And wouldn't an investigation do more harm than good— interrupt the work, disrupt the unit, ruin a good man's career?

Mason and Messing stand with him. Together they calm the men down, and the crisis passes.

In April, when the snows melt and reinforcements arrive, the Allies begin their spring offensive, slowing down the wave of spies and giving Jocko time to write three reflective V-mails. He's been in the army two and a half years now and there's no end in sight. He wants to get married and have a family. He wants to marry his Italian girlfriend, but she has ties she can't seem to escape. Maybe he'll accept some foreign assignment and establish a life overseas somewhere. And he's obsessed with his new wrist-watch, which he checked at least fifty times today. Nobody from the normal world could imagine how precious a luxury a wristwatch can be to a soldier. And soon the Italian intelligence man he lives with will come in and say *buona notte* and he will say *buona notte* back and he will go to bed and sleep it off. How strange life is!

> Do you remember, George, how I was the bookworm who never got out of the library? Well, I've become without exaggeration, jailer, judge, interrogator, politician, man-hunter. This is the sober truth, and seems in no way romantic or out-of-the-ordinary to me.

Then the Allied troops break through the Fascist battle lines and chase the enemy across the Po Valley, unleashing a new wave of enemy spies. In the space of a few weeks, Jocko catches twenty of them single-handed, one of the rare exploits he includes in his letters to me, though he won't let the paragraph end without a dash of his usual chagrin. "They were a poor lot really, nobody really smart or able, nothing Hollywoodish about the matter."

But there was, he admits, "a certain excitement in it."

13

•

TRIESTE

That year the Allies carved up the world, drawing borders that quickly took on the feel of fresh battle lines. The troops were still stuck in the Italian snow when Churchill and Roosevelt met Stalin at Yalta and agreed to divide Germany and let Stalin "supervise" the Polish elections. But Stalin didn't stop there. In March, his troops deposed the generals who were running Romania and imposed a Communist government. A month later, when Churchill sent troops to Greece to help put down a Communist insurrection, he started seizing power in Bulgaria and Hungary. Then British and American troops pushed through northeast Italy and came up against Stalin's ally in Yugoslavia, Josip Tito. As the defeated German and Croat troops tried to surrender to the British and Americans, certain they would get better treatment from them, Tito's soldiers started gunning them down. Since Tito and Stalin were members of the Allied coalition, the British and Americans could only watch in horror and disgust. A replay of Stalin's ruthless pause at Warsaw, this massacre was done for similar reasons—so Tito could eliminate his enemies and secure control of the city of Trieste, a prize port long disputed by Italy and Yugoslavia. Finally General George S. Patton moved in with three troop divisions to stop the slaughter. Tito responded with five troop divisions. Suddenly hundreds of thousands of armed "allies" were bristling up against each other, coming so close to violence that when Stalin finally ordered Tito to stand down in June 1945, he said it was because he didn't want "to begin the Third World War over the Trieste question."

And my father? As the war ended and the troops trudged forward to occupy Europe, he got in his Jeep and drove past the confused German soldiers shuffling along the sides of the roads to a new post heading up the CIC office in Bologna, which had a reputation as—I'm still astonished to see that he actually wrote this down—"the oral sex center of Italy." The other memories he dog-eared in his mind were along the same lines. Spingarn bounced back, suddenly just as cheerful and friendly as before. He went with Mason for a holiday in Venice, a moment of relaxation so welcome that he shocked Mason by springing for a lobster dinner. And one day a young German officer knocked on Jocko's door and asked for a moment's indulgence. "I have a story and a request," he said.

That was an intriguing approach. So he nodded the man to a chair and told him to tell the story.

It was like the plot for a romantic movie. A few months earlier, the German officer said, he had deserted under fire and escaped into a hidden mountain village where a beautiful Italian girl gave him refuge. As the days passed and spring came to the mountains, he helped her in the fields and fixed things around the house. And little by little, hidden from the world, they fell in love.

But now their secret was out and the villagers were mad for revenge. They punished the girl for loving a German by shaving off her beautiful black hair. The children threw rocks at her, the old women spat as she passed.

So that was the story. And his request was simple—permission to go save her. He promised to return in fifteen days. "I give you the word of a German officer," he said.

The word of a German officer? Just a few months after American troops liberated Auschwitz?

But my father always seemed proud of his answer, even a bit sentimental about this moment when the vigilance of war relaxed and forgiveness seemed possible. And he always ended the story the same way:

Exactly fifteen days later, the German officer came back.

But the sunny mood didn't last. As Messing and Mason and many of his other friends headed home in relief, taking up civilian lives again, my father chose to stay. The work was important and he had no other obligations, after all. His mission shifted from catching spies to arresting Nazi officials. "Denazification," they called it. The idea was to destroy the Nazi system by putting its leaders into internment camps while building new governments and legal systems. After more than a decade of Nazi power, it was a massive job. Within the next ten months, in Germany alone, CIC agents arrested 120,000 Nazi officials. They turned the Dachau concentration camp into an internment camp.

My father was sent to Austria, which posed special problems. Ever since Hitler arrived to cheering crowds during the *Anschluss* of 1938, driving into Vienna in an open convertible, it had been the most pliant of Germany's conquests. Seven years later, Nazis ran everything from the parliament to the sewers. When the Allies summarily fired all party members from their government jobs, it took until December just to find enough teachers to open the schools. And the country was also teeming with refugees, hundreds of thousands of them, traumatized survivors of labor camps and concentration camps, deserters from the Russian Army and Jews desperate to emigrate. Posted to a ski resort named Zell am See, a charming little town set on a large mountain lake with spectacular streams and forests in all directions, my father must have started his mission with another blast of chagrin. But Zell am See attracted so many fleeing Nazis that he began arresting them at the rate of fifty a month, taking particular pride in the capture of the local Gestapo chief, "a thoroughly brutal person." The job had many difficulties: records were missing, evidence was spotty, people disagreed, informers were often trying to settle old scores or scheming for profit, and it was sometimes impossible to distinguish the people who merely compromised from the true Nazi believers. At other times, it was shockingly easy. One day, he arrested an SS captain and took him back for the usual interrogations, only to sit there stunned as the man confessed to participating in the liquidation of the Warsaw Ghetto, telling the story in detail and even showing him pictures.

Darker problems swirled around him. Years later, for example, during his trial for war crimes that included the torture and murder of over one hundred people, a Nazi named Robert Jan Verbelen testified that he started working for the Americans as a bartender at an officer's club in Zell am See right after the war and shifted over to a job spying for the CIC soon afterward. Although Verbelen managed to conceal his crimes for years and may not have worked directly for my father—he said in a later interview that he didn't actually start working for the CIC until May of 1946, long after my father left the area—he's an early example of the moral risks the CIC men were beginning to run, foreshadowing the intelligence scandals to come.

Once again, my father had the consolation of loyal friends. Bob Cunningham came up from his CIC post in Italy and later remembered him "arresting Nazis left and right." His supervisor was a Harvard lawyer named Gerry Weber, who took a picture of him on the highest mountain peak in Austria with one hand shoved in his pocket and the other holding a lit cigarette, an expression of deep concern on his face. "A man with his head in the clouds but both feet planted firmly on the ground, in the best American tradition," Weber captioned it. And other comrades-in-arms were making arrests and building cases all over Europe, climaxing in the trials that began that November in Nuremberg. As news of the Nazi horrors spread, my father threw himself into the job, working so hard that later the Austrian Ministry of the Interior cited Zell am See as the most thoroughly denazified county in all of Austria.

Years later, recalling this period, my father told one story more often than any other. He had a scrapbook of magazine pictures from Auschwitz and Buchenwald and the other concentration camps. Just before sending a Nazi to an internment camp, he said, he would make the man sit down in his office and leaf through the scrapbook.

> I had come to hate the Nazi system, and I mean hate it emotionally as well as intellectually. I was glad to participate in the punishment of these men whose punishment was mild in contrast to the brutalities the Nazis had perpetrated on millions of people.

Sometimes the Austrians complained, he remembered. They said the Americans came to bring freedom but instead brought internment camps.

He would always give them the same answer: "We didn't come to bring freedom to people who participated in the tyrannical destruction of freedom, but to their victims."

There were other consolations. In Zell am See he fell in love again, an odd kind of postwar romance that began at a mansion estate in the country where he spotted an Austrian girl so lovely he felt almost breathless. Forty years later he didn't recall her name but still remembered that first glimpse. She was tall, slender, blue-eyed and golden haired, with a perfect fair complexion and "a slight sprinkling of almost imperceptible freckles about the nose." Surprised when she spoke to him in English and delighted to learn that her mother was an American, he was already smitten by the time she told him she was married to an SS officer and had two children. But her husband was lost in the postwar chaos, possibly in a prison camp, and her children were in Vienna with their grandparents. Maybe she could get a divorce, maybe they could start a new life together.

Then his successes in Zell am See led to a new job as head of the CIC office in Salzburg, where the work got even more complicated. Best known as the birthplace of Mozart, Salzburg had become so popular a Nazi hideout that the locals would joke: "If you're not a Nazi, what are you doing here?" It was also popular with Yugoslav exiles, who started showing up in Salzburg that winter, trying to buy military equipment and trading in the black market. Although most tried to pass themselves off as partisans of Vladko Macek, a heroic Yugoslav politician who refused to make a deal with Hitler, many were members of the Ustaša, a brutal army of Croatian Fascists that massacred some four hundred thousand Serbs, Gypsies and Jews during the war—an intrigue that would soon become the center of my father's life.

Salzburg was also the headquarters of the Russian repatriation unit and the largest refugee camp in Austria, which would turn into one of the primary irritants of the long occupation. Since few Eastern bloc refugees wanted to go back and the Soviets were given to kidnapping and other forms of forced return, small disputes kept bubbling over into political showdowns. CIA officer and historian Harry Rositzke identified this as one of the emotional sources of the Cold War. "In Berlin, Vienna, and along the frontiers of the occupation zones, we saw firsthand the brutal handling of repatriated Soviet citizens and the wholesale arrest and deportation of countless Germans." Finally the Americans refused to cooperate at all, a decision which became personal for my father when one of his men married an escaped Russian slave laborer named Valia. Forced to work in Nazi coal mines since she was sixteen, she now lived in fear of being sent to a Russian prison for collaboration. One day, a high-ranking Soviet officer came into his office to protest the new American policy. As he described it later, the man arrived with an armed assistant and immediately started shouting and huffing around the office in a bullying, overbearing manner, typical of the Soviet style. "After being on the receiving end of this for a short while, I told him to get out of my office immediately and not come back. If he didn't leave immediately, I'd have the MPs throw him out."

The officer's tone changed instantly. There was no need for harshness, he pleaded. This could be solved in a reasonable way. It was an important lesson for my father, one he tried to pass on to me.

> All subsequent experience has convinced me that you can
> deal with the Communists and the Nazis of this world—and
> all bullyboy types—only from a position of strength. Their
> basic human philosophy, if you can call it human, is that of
> the bully—despise and abuse weakness, defer to strength.

Through all this, he continued to drive the icy mountain roads to Zell am See to romance the girl who took his breath away. Although the "bullyboy types" aroused his disgust, he was never a harsh judge of the people who got pulled along in their wake. Like the Italians, they were just trying to survive. Their weaknesses were human weaknesses. If she loved the SS officer at first, it was human, no different the good fellows at the *komaradschafthaus*. But his detachment also gave him the ability to see things with a cool clarity that could be disorienting—like the day she said that if they were to get married, she could send her children off to live with relatives. For a moment he thought of Medea killing her children.

But he still loved her, he decided. And he would marry her if she could get free.

Then a man named Burt Lifschultz arrived with an intriguing proposition: despite the dissolution of the OSS in September 1945 and hostility from Truman and newspaper headlines warning of a "super Gestapo agency," there were enough people in power who understood that America was going to need some kind of intelligence group in the years to come. So far the best they'd been able to come up with was a vague new outfit called the Strategic Services Unit. The future was uncertain and the pay wasn't much and once again there would be no chance for glory or recognition, but it was important work and they wanted him to be a part of it.

By this point, most soldiers just wanted to go home. Within nine months of V-day, 87 percent of the OSS quit.

Jocko said yes. That December, the U.S. Army promoted him to first lieutenant and discharged him, and immediately he joined the Austrian station of the SSU with the rank of GS-12, the civilian equivalent of a captain. His salary was $5,200 a year. His first posting would be to Trieste.

Two days later, just before Christmas, he sat down in an office somewhere in Vienna and typed his last letter to George Chisler. Freed from wartime restrictions and caught in a reflective mood, he told the story of his war from his arrival in Italy and his friendship with de Franchis to the grim days of the rebellion against Spingarn, who was so "brutal, harsh and severe" that some of the fellows could not forgive him even now. He revealed the feelings he held back before, like his shame at the deference of officers who outranked him and his fear of being "unmasked" in some embarrassing situation. He admitted that for all his denials, the work was constant. They never stopped moving. There were always new problems needing solutions. They were always under pressure. A stark reminder of how deeply he had taken to heart the conversion described in the long confession he'd written years ago in Berlin, this letter drove home just how well he'd learned to keep it all inside. Gone is the boy who gushed over Katharine Hepburn in *The Little Minister*, the boy who tried to dance the gypsy dance. The changes have reached into the rhythms of his prose. "Looking back at my peculiar war years," he mused, "I feel older than the three years would have normally caused, sadder and very tired."

> I drank hard, played poker and shot craps, made love indiscriminately like all soldiers do, fell in love once with a Milanese who lived in Centurano near Caserta. In three years I have hardly read a book, and feel now almost too restless to spend a single evening at home.

Then the V-mails to Chisler stop.

Before my father disappears into the secret world, I have a few last glimpses of him. First there's the version he gave in his last autobiographical letter to me. Driving down from Vienna with no special training or preparation, his title "Chief of Base," his mission to gather "positive intelligence" about the Yugoslavs, he arrived at a pivotal spot in a pivotal time. The streets of Trieste were clogged with massive demonstrations by Yugoslavs and Italians, both struggling for control of the city. There were ominous stabbings and shootings. Just a month later, Stalin gave his first major postwar speech to the Supreme Soviet, saying that the war had been caused by the internal contradictions of capitalism and suggesting that future wars were inevitable given "the present capitalist conditions of the development of the world economy." Truman and Churchill took this as a menacing restatement of the old Communist internationalism, which seems to be pretty much how Stalin intended it. Just a few weeks after Stalin's speech, George F. Kennan sent his famous long telegram from Moscow outlining the containment policy: since the Soviet Union seemed bent on expansion, he argued, we would have to counter their moves in a grand global chess game. A few weeks after that, Churchill captured the moment in a memorable phrase: "From Stettin in the Baltic to Trieste in the Adriatic, an iron curtain has descended across the Continent."

My father started working out of a comfortable downtown apartment building. He shared his rooms with an Army counterintelligence officer and a Croat assistant who called himself Bob Perry, formerly one of the mountain warriors who fought the Nazis with a Yugoslav guerrilla leader named Draza Mihailovic. His staff included an Army corporal who served as clerk, a couple of Slovenes and "a murderous-looking Serb named Steve." Again, he developed a close friendship with one of his sources, a former Serb police officer. Almost every day, he sent a report to the SSU station in Vienna about politics, smugglers, and spies.

Beyond this, he told me a single story. Just as he was settling into his office, he said, an American Army major asked Perry to help handle a currency transaction. The major was in Trieste doing counterintelligence work for Jim Angleton, the SSU chief in Rome. As it happens, my father and Angleton had worked different ends of the Carla Costa case, when Angleton directed the background investigation in Rome and later confirmed the work my father and the Gordons had done through his own interrogations.

Now he was a particulary powerful SSU man, spending millions of dollars to fight off an alarmingly strong election campaign by the Italian Communist Party. But Perry thought the Army major was up to some kind of scam and went to my father, uncomfortable and unhappy. Should he help the major anyway?

Absolutely not, my father told him.

But late that night, he woke up and saw Perry getting dressed in the kind of outdoor clothes guerrilla fighters wore. "What are you doing?" he asked.

"I'm going to Yugoslavia to join Mihailovic," Perry said.

Rather than getting caught up in a conflict between my father and Angleton, Perry was choosing an almost certain death. At the time, Mihailovic was fighting a last and futile stand against Tito. More to the point, perhaps, Perry was choosing to join an anti-Communist leader the Americans snubbed (some say betrayed) in the name of their wartime alliance with Stalin. It was a grand gesture of high principle.

Somehow my father managed to talk him out it. The next day, he called a meeting with the Army major. "I have to report this to Jim Angleton," he told them.

"Then I will have to go to Rome with you," the major responded, his tone formal.

It was a showdown. Years later, when he told me this story, my father chuckled and said he supposed there was a possibility he might not make it to Rome. After all, it was his expressed intention to turn the man in for corrupt behavior. If he was really corrupt, if they were armed and alone together on lawless highways, who knows what could happen?

But there was no turning back. They piled into a Jeep together and drove east, talking in a courtly and civilized manner all the way. And just before they reached Rome, the major turned to speak. "John, I hope you won't make this too rough on me."

"I'll do what I have to do," my father said.

At SSU headquarters, they filed into Angleton's office. In his memoir, CIA chief Richard Helms painted the legendary agent well: "His not entirely coincidental resemblance to T. S. Eliot was intensified by a European wardrobe, studious manner, heavy glasses, and lifelong interest in poetry." After my father explained the situation, Angleton sent the major out of the room and told him that while the crime was serious, the major was doing very important work and it would take a while to replace him.

With that, my father and the major got back in the Jeep and went back to their jobs—and Angleton never replaced the major.

When he got to this part of the story, my father laughed in a way that suggested Angleton knew something he didn't. Maybe the currency irregularity was part of some other scheme, maybe the major was too important to sacrifice for reasons of petty dishonesty. But I can't help thinking that the anecdote was a metaphor for the increased moral and political complexity of the world he had entered. And much later, in Box 90 of the SSU records in the National Archives, in a "name file" labeled Richardson, I found a hint of just how complex it was. It begins with a coded cable sent by General John Lee of Army Headquarters on December 28. The previous SSU chief in Trieste confined himself to reports on the Yugoslav order of battle, he says. But this new fellow is stirring up trouble. "It appears that Richardson is more directly contacting in Trieste Jugoslav refugees with interests and contacts in Jugoslavia proper."

Just as Yugoslavian refugees had flooded into Salzburg, they were flooding into Trieste. Some were members of Vladko Macek's Croatian Peasant Party, some members of the Fascist Ustaša. Apparently my father was putting the lessons of Salzburg to use, trying to develop aggressive new sources. But the general was worried that Tito's secret service—the dreaded Ozna, infamous for its ruthless campaign of political assassinations—was looking for an opening to use against the Americans and British. "It is known that Ozna has largely penetrated these refugees," he continued. "Primary function in this area of Ozna is believed to be collecting information proving collusion between anti-Tito elements and Allied Intelligence organizations."

This complaint sparked a flurry of coded cables from Paris to Rome to Vienna. Soon reassuring words come from SSU headquarters about Richardson's "excellent record, his ability to produce intelligence, his carefulness in security matters." One of his supervisors said the Army bureaucrats were probably just alarmed by the "fairly hot intelligence Richardson was giving corps." It seems my father was in contact with Macek supporters and also following the Crusaders, a group of former Ustaša soldiers who were busy planning a war of resistance inside Yugoslavia. At this time, in a shocking bit of tactical expediency, Macek was secretely conducting peace talks with leaders of the Ustaša with the aim of uniting the opposition. Although some historians have accused Jim Angleton of trying to cover up his contacts with the Ustaša, the cable that appears in my father's file makes it clear he was following the peace talks closely and was trusted enough to speak frankly with a top Ustaša general named Vilko Pečnikar. "Pečnikar against any infiltration resistance groups without Allied support due to lack of arms and materials," he wrote. "Pečnikar has great faith in subordinate Col. Ivan Babić."

Soon afterward my father sent a cable saying that the Peasant Party and the Crusaders had united under Babić's leadership. They were planning an uprising against Tito in the spring and wanted to make a formal request for Allied support.

Their request went all the way to the head of the SSU in Washington, General John Magruder. After considering the geopolitical implications, he ordered Angleton and my father not to give the rebels any help.

> Regardless personal reliability Macek, his operations envisage revolution against government recognized by U.S. Our support his service could be construed by him or Tito as support Croat movement. In either case publicity could be disastrous.

At that point, my father seems to have turned his attention to arms smuggling and Ozna. The SSU files give occasional glimpses of how he spent his time:

In June, a man named Modesto Limone gave him "outstanding reports" on the black market.

In July, he had to retire a spy named Giovannoni after he was approached by a suspected Ozna agent who knew him before the war.

In May, he accepted an extensive report on a slender young woman named Franca with light chestnut hair and a reputation for murder— apparently she led the Ozna network in Italy, though few people had ever seen her. Headquarters also sent a dossier on the head of the local arms smuggling business, a man named Max Samuels D'Acquaviva. "Contact him and shadow his activities," they ordered.

But my father never mentioned the Crusaders, never mentioned the Peasant Party, never mentioned the Ustaša. Their guerrilla war against Tito started on schedule that spring and sputtered on for another three years, but he never said a word. He brought up the street violence only to insist that it was never as bad as the newspapers made it sound—and yes, the bodyguard who protected his policeman friend was shot and killed one night just around the corner from his office, but it seemed random. They certainly didn't live in fear.

The closest he got to drama was the last paragraph of his last autobiographical letter:

> But I'll always remember the Yugoslav machine guns and artillery up in the hills above the city, trained on the only road out of town, and the special and personal sense of pleasure we all felt when Truman sent one of Uncle Sam's great gray battleships into the harbor to anchor and point their guns toward Yugoslavia.

With that, his letters stopped.

When the first group of Westerners entered Vienna in June 1945, they found more than one fifth of the buildings badly damaged or destroyed, debris piled in heaps on the sidewalks and most of the windows covered with rough planks or oilcloth. People wandered the wreckage collecting scraps of wood for cooking and heating. There were only three working ambulances. The sewage system was broken everywhere, half the telephones were out and most of the electrical grid was down. Russian soldiers were busily stripping the city of furniture and factory machinery and telegraph equipment along with thousands of watches, radios, telephones, cars, tools and even park benches. The Austrian government later estimated that the Soviets looted 20 percent of Austrian agriculture and industry. They even cut the leather out of the seats of chairs. And they decreed that all the imported food from Hungary and Romania and Yugoslavia would have to flow east to feed their own starving people, leaving the Austrians with less and less to eat. Bread spread with lard became a luxury. The argument over how to feed them dragged on for months, but finally the Allies divided the city into five sectors with control of the heart of the city—the Innere Stadt—passing from one Allied power to the next on a monthly rotation. The Soviets set up their headquarters in the Hotel Imperial, the biggest and most magnificent hotel in Vienna. The British took the Hotel Sacher, run by a cigar-smoking eccentric who would later become world famous for her *torte*. The French took the Hotel de France, which had the best dining room. The Americans took the Hotel Bristol, its lobby dominated by a vast staircase and an immense painting of Emperor Franz Josef. Soon a few restaurants reopened and by early October 1945, the famous Vienna Opera tried to jump-start civilian life with a performance of *Fidelio*. Courtesy of an American general, someone found five thousand baby bottle nipples for the grateful mothers of Vienna. The hopeful symbol of the city was "four men in a jeep"—Innere Stadt patrol vehicles manned by soldiers from each of the four powers.

But the hope died fast. The Soviets demanded half-ownership of the Austrian oil fields, then refused to share the fuel with the other zones. They used the cover of war reparations to seize Austrian businesses and farms. They tried to take over the largest shipping company. Locals joked that the Russians rotated troops to give a fresh batch of soldiers a chance to loot, or that the immense statue they put up in Schwarzenbergplatz was really a tribute to the Unknown Plunderer. They made travel difficult too, refusing to let anyone cross the Soviet zone without a pass. They even refused passes to the American and British teams searching for war dead, stretching a job estimated to take eight weeks to more than two years. But they insisted on free access to the refugee camps, where they continued to "repatriate" citizens so unwilling they had to be dragged away at gunpoint.

A year into the occupation, most people were still dressed in patched clothes and living on potatoes flavored with pea paste. Nine thousand had tuberculosis.

By the time my father arrived—I believe he started in early 1947 as deputy to a station chief named Al Ulmer—the city was crawling with spies. The KGB and the GRU were working out of the top floors of the Hotel Imperial and the SSU had morphed into the Central Intelligence Group, hiding its agents in offices with innocuous names like "Plans and Review Section." The U.S. Army had G-2 and the CIC, and the French and British had spies too. Then there were freelancers like the Black Brigades, a pro-Soviet group that trained factory workers in street fighting, espionage and sabotage. At the lowest rung, there were hungry Austrians and refugees who would bolt into U.S. government offices or approach soldiers at the refugee centers, offering information for money or a visa or even cigarettes. Sometimes it was a great deal. A Soviet army deserter named Lev Durow told them that Russia was laying the groundwork for an invasion of Iran by sending teams into the desert to drill for oil in secret and running its arms factories around the clock, a tip that helped raise the alert about Soviet expansionism. A part-time spy named Rudolf Schwab tipped them off to a plot to smuggle explosives into the British zone.

At first, like other agents, my father spent a lot of time interviewing defectors and sympathetic Austrians, trading soap and coffee for fresh glimpses of Hungary and Czechoslovakia. He ran names by the local police and friendly intelligence agencies. He tried to gather information on the explosion of smuggling, which Washington studied as a kind of X ray of the economic situation across the Iron Curtain. He helped shore up the police force, the same kind of "nation-building" work he'd done in Italy. He spent countless hours reading the dossier files on local figures, the heart of any intelligence service. By 1950, for example, the Austrian CIC had nearly 100,000 individual dossiers filled with details about voting records and drinking habits and sex lives. He researched the Austrian Communist Party, learning the difference between *ortsgruppe* and *bezirksorganisation*. When he and his friends caught their clerks and cleaning ladies spying for the Russians, they tried to "double" them, pressuring them to take back false or useless information so they could spy on the spies who were spying on them. When the clerks and cleaning ladies wept and said they had family members trapped in the Soviet zone, he tried not to let it upset him too much and kept his eye open for "confusion agents," ordinary citizens tricked up to look like spies and waste investigation time. He became an expert on the bars and back alleys of the Innere Stadt.

Some nights, he went out carrying a gun.

After a few months, my father took over as station chief. Now he was thirty-four years old, an important man with heavy responsibilities. Working out of the same massive old building as the Allied High Commissioner, he met regularly with Austrian authorities and military intelligence. He supervised ten case officers and two bases, one in Salzburg and one in Linz, plus about twenty secretaries and the reports officers who compiled and distilled the raw intelligence. And with the new job came a pleasant villa with a large garden and a butler named Fritz, who learned to make martinis and serve them very cold.

War teaches people to take fierce solace in whatever pleasures they can find, and the people of Vienna took this lesson to extremes. Four months after liberation, despite bomb damage to the Opera house, the Vienna Philharmonic struck up the Polonaise. People cut up old clothes to make elaborate costumes for the annual Artists' Ball. They paid twenty-five dollars for silk stockings on the black market, almost two months' salary for the average worker. Within a year, there were forty theaters and two opera companies playing to packed houses every night. At the variety houses, audiences watched fashion shows and historical tableaus and cried at sentimental songs about mother love and poor Vienna. "The twentieth century has no time for romance or happiness," went one popular lyric. "Just look around, look at how your city has changed—rubble, ruin, misery, and murder."

Then the chorus crooned: "But music from Vienna is still the most beautiful thing."

My father absorbed this spirit and made it part of his philosophy. Mixing work with pleasure, he went to the balls to meet Austrian conservatives and hung out in nightclubs to cultivate the socialists. He began going to diplomatic parties dressed like a lord in gleaming black tie. He discovered he had a gift for talking senior brass into anything the Station needed and didn't scrimp on the luxuries—like the elegant "Joe" houses for whatever Joe might happen by, including one villa so grand it had a full-sized organ built above its fireplace. He spent time at the press club, drinking martinis and playing poker. He dated the young ladies of the office, first a woman named Mary Whittaker and then someone named Lee Keith. He also put the moves on a young reports officer named Sue Darling, who told me years later that the old man was "quite macho" in those days. "He checked out the field," she said. "Every one of us was checked out."

By now, he was the old man in the group. He enjoyed weaving a bit of mystery about himself, taking some pleasure in what one of his deputies called "oriental inscrutability." Asked why he was born in Rangoon, he said "That's where my mother was." Years later, one of his old friends told me that Vienna was where he became "a master at talking without saying anything."

For a period, his old depressions came back to plague him. At one point he slipped into a funk so dark, it was almost like the college days all over again. The stubborn streak he'd shown in the Beech episode resurfaced too. When the Central Intelligence Group became the CIA and an order went out that all officers had to take a lie detector test, he refused. It was just a security measure, but he wouldn't budge. If they didn't trust him after all his years of service, he said, they could damn well fire him. Finally Jim Angleton said he could skip it, an exemption that apparently made him unique in CIA history.

That June, the Truman Doctrine made containment the official American policy. Now moving through the Soviet zone became more and more difficult. To get to the nearest airfields, you had to drive through a narrow corridor in Soviet territory. From the British zone to the American zone, you had to take a closed train. Drunk G.I.s who wandered into So-viet areas often ended up spending months in Soviet jails. One American who accidentally stepped across the wrong street was detained for fifteen hours, slapped around and released on the outskirts of town without his papers.

The spies went into overdrive. In a memoir called *Partners at the Creation*, a CIA officer named James Critchfield remembered Eastern European politicians arriving in flight from Soviet crackdowns and Zionist organizations like the Stern Gang smuggling hundreds of Jews in and out of a huge refugee camp right in the middle of the city. "Shootouts, occasional kidnappings of Austrian politicians, the arrests of agents, and occasional mysterious murders were typical."

Sometimes, it was sleazy. In the CIC records in the National Archives, I found the story of a hotel clerk named Josef Hutter. To get the money to cure a case of gonorrhea, he agreed to steal a U.S. army gas mask. Another file told the story of Anna Wukowitz, a twenty-two-year-old Austrian girl who was ordered by her boyfriend to seduce a British soldier who had access to military plans. They got as far as blood tests and a marriage application before the CIC caught them.

There was also pressure to use Nazis and other war criminals as agents. That year, for example, the CIC set up an escape route called the Ratline to smuggle important Russian and East European refugees to Rome, where they got visas to South America from a Croatian war criminal named Krunoslav Dragonovic. A few years later the CIC used the Ratline to smuggle out Klaus Barbie, an infamous Nazi who tortured and killed the leader of the French Resistance and sent thousands to the death camps.

The CIA used the Ratline too.

But my father's friends said that he resisted using Nazis as agents, and I found one piece of direct evidence from this period to support this—a letter from his office in April 1947 about whether to hire a former SS man named Otto Abrecht von Bolschwing:

> After considering the information on Subject provided by headquarters, together with Heidelberg's reply to our inquiry, we have decided not to use Subject in any capacity. No approach will be made to him.

But the pressures increased. In February 1948, Communists took over the Czechoslovakian government with a sudden coup. In March, the Soviets expressed their anger over the division of Germany—they wanted it all—by withdrawing from the commission that ran Berlin and slowing down the trains carrying Western troops into the city, an ominous prelude to the Berlin Blockade. In Italy and France, Greece and Iran, China and Vietnam, Communist parties and revolutionary armies were pushing to power.

Alarmed, Truman ordered the CIA to establish a covert action arm to fight back. This was the Office of Policy Coordination, the OPC. Using contacts supplied by a former German spymaster named General Reinhard Gehlen, an ex-OSS man named Frank Wisner took another step across the moral line by building a secret army of former Nazis and Nazi collaborators in hopes of giving the Soviet Union a taste of revolution at home. He smuggled more than three hundred of these men into the United States for training in intelligence gathering, covert warfare and even assassination. Later he hired three former SS men as full-time agents and financed a private intelligence group called the "Org," which grew into modern Germany's intelligence service but also put on its payroll at least a hundred ex-Nazis.

My father was friendly with Wisner. He was also close to the head of the OPC's Austrian office, a handsome Canadian named Lochland Campbell. He was certainly aware of what OPC was doing in Austria, scheming to influence various newspapers and political parties as well as coordinating with the Salzburg branch of the Org and running Radio Free Europe, the OPC's one great success. But it's my impression that my father stayed fairly hands-off and felt that Wisner's team had a tendency toward ambitious and unrealistic plans. A deputy chief named Bill Hood told me that one time one of the OPC guys came back from Washington all revved up with a plan to "bring Hungary to its knees" by buying the country's largest ball-bearing factory. They thought he was being silly. And a case officer named Jean Nater told me that the aggressive CIA men in Germany sometimes dismissed my father's young staff with a scornful label: The Vienna Choir Boys.

One day that year, Sue Darling came back to the CIA office talking about a charming Austrian she had met on the train, a little old lady with gray hair who'd been traveling in the Soviet zone and knew an amazing amount about their reconstruction efforts. Apparently she worked for the ministry of economic planning. Maybe she'd be useful.

A few days later, the Soviets kidnapped the woman—her name was Margarethe Ottilinger—in broad daylight.

That shocked them, Darling told me.

Shortly afterward, an American diplomat named John Erhardt sent an alarmed cable to Washington. "Series of recent arrests and kidnappings by Soviets have considerably increased feelings of insecurity among Austrian officials and people generally." The latest incidents included an Austrian who jumped to his death from the window of a Soviet police office and an American aid official named Irving Ross who was beaten to death by Soviet soldiers. There was also a fatal scuffle on the Mozart Special train when two Soviet officers tried to push their way past an American sergeant named Shirley B. Dixon, who ordered them to stop and shot them when they refused, an incident that led to furious protests from the Soviets and even more tension. While most of these events stemmed from actual or suspected cases of espionage, Erhardt said, the Soviets seemed to relish the "secondary advantages"—fear and subservience.

One night when my father was working late, all this became very personal. The problem started up in Romania, where George Mason was back in intelligence work after a brief sojourn in civilian life. The Romanian Communists were so hostile they painted a bull's-eye on the wall around his house and used it for target practice. Then Soviet troops began to mass on the border, and Mason went out to the countryside to take pictures, only to be stopped by police who seized his film and held him in a cell while they sent his film to be developed. When it came back a few days later with nothing but innocent shots of lovely fall foliage, they grudgingly released him—along with the undeveloped roll of Russian jets stashed in the cuff of his pants. That's when he decided things were getting too hot and sent his wife down to Vienna.

She made it all the way through Romania and into the Russian zone of Austria. At a Russian checkpoint just outside Vienna, the guards finally stopped her.

My father was working at his desk that night and got the urgent call: Louise Mason was in Russian hands. He rushed straight out to the checkpoint and demanded to see her, warning of terrible consequences if they didn't produce her immediately, applying his maxim about dealing with bullies from a position of strength.

They released her into his custody.

Then the Soviets launched the Berlin Blockade, cutting all train and road traffic, and my father began working on a plan to resist a similar blockade in Vienna. Under the code name "Squirrel Cage," he designed drop points and distribution networks. But there was a crucial strategic difference. In Berlin, the Western powers were able to feed the people by launching the Berlin Airlift. In Vienna, the airports were on the other side of the Soviet zone.

During this tense period, a case officer brought my father an elaborate report about a Soviet plan to invade Austria. It was from a good source in the Austrian foreign affairs ministry and looked solid, so he took it to John Erhardt, who became very somber and said it was too important to share with G-2 or the British—they should rush it to CIA headquarters right away.

When the report reached Washington, limousines raced from building to building, stocking crisis meetings with Very Important Persons. A senior CIA officer rushed across the Atlantic to interrogate the Austrian source in person. Soon word of the report reached the Washington head of military intelligence, who immediately sent a cable to the Vienna G-2 chief asking for more information. And that man told his staff he was sure his close friend John Richardson would share it with him.

But Richardson didn't. He obeyed his orders and kept his mouth shut.

In the end, the source confessed that his report was a fake and the G-2 chief paid back his embarrassing moment with interest, teasing my father for getting so excited over a forgery. That made the story a candidate for my father's Tales of Chagrin, which is the only reason I know it. But though his subsequent situation reports seem to have advocated a calm approach, arguing that Russia was unlikely to risk war by mounting a blockade, the city remained tense. The Red Army could sweep through at any time, people said. And the U.S. Army wouldn't be able to resist. It would lead to a general war, they were sure of it.

Inevitably, the pressures of occupation pushed my father and his friends inward. Fritz learned to mix his martinis in liters and keep them in jugs in the refrigerator, and they spent many nights talking and playing cards and playing charades, always more comfortable with one another than with anyone from the outside world. When you got my father in the right mood, his friends told me, he would play the piano and laugh and even crack jokes. He teased one friend about being a real Don Giovanni, one step above the gates of hell. Another remembered how eagerly he threw himself into charades, once trying to signal the word "Pharaoh" by feverishly whipping imaginary slaves. But his favorite activity was marathon conversation. One case officer and his wife remembered him coming over for dinner and talking until their baby daughter woke up in the morning—mostly about Marcus Aurelius, his favorite Stoic philosopher. Right here in Austria, he told them, Marcus Aurelius spent his life fighting the barbarians. Right here, he died and was buried.

In the middle of telling the story, this case officer went into his home office and came back with a copy of the *Meditations* of Marcus Aurelius. It was a gift from my father, his signature on the flyleaf. It fell open to a random page:

"I do my duty. Other things do not trouble me."

But soon my father would find a consolation beyond philosophy.

15

●

MARTINIS, POKER, CIGARETTES

In the fall of 1949, a striking young American named Eleanore Koch arrived in Vienna. She had full lips and dark hair and an air of innocent fun that seemed like a breath of hope and home, and she quickly fell into the expatriate world of journalists and embassy officers and parties that centered around the International Press Club. Although she was planning to stay only a month, having come to visit her brother Dean, she jumped at the chance to give lectures at military hospitals on speech pathology, the subject of her freshly minted master's degree from Columbia University. There were a lot of wounded soldiers who needed to learn how to speak again. Suddenly she was flying into Berlin in the middle of the airlift.

It was all so exciting she decided to stay for a while and got a job working in the U.S. Army's Austrian Youth Program in Linz. To drive through the Soviet zone, she bought a little-used Topalino, an Italian car so small that two strong men could lift it right out of a tight parking space.

Dean was a high-ranking military intelligence officer. One night in November, he took her to an elegant cocktail party at the home of a *New*

York Times reporter named John McCormick. The guests were a romantic group of war-hardened men, from a battlefield reporter for the *Herald Tribune* to a political correspondent for CBS to a former music critic who had climbed to the top ranks of military intelligence. Gordon Messing was there too, having joined my father's team after a brief stint as a college professor. Dean brought his wife, Nancy, a lovely but scatterbrained young woman with secret socialist leanings and a taste for unconventional behavior. Back in New York, they had lived together in sin until he finally convinced her to marry him. Eleanore's date was Si Bourgin, a tall *Time* magazine reporter with horn-rimmed glasses and a merry smile.

In the spirit of Vienna's defiant elegance, they were all dressed as extravagantly as possible. Fifty years later, Eleanore would still remember her outfit: "I had on a black dress with a lace top. And an orchid at my waist that Si bought me."

Jocko Richardson didn't make much of an impression at first. He was a bit plump and quite a bit older. People were teasing him about his paunch but they seemed to respect him and even defer to him. But with her he was very modest, mostly talking about the trip he just took home, where his friend George Chisler dropped him off on one side of the Sierra Nevadas and he spent twelve days hiking down the other, fishing for his food—the beautiful mountains he had dreamed about during the long war.

When Si caught her eye and said it was time to go to dinner, Jocko invited himself along. Bourgin found it a bit surprising but didn't object.

They took Jocko's car, which was waiting at the curb. The dinner went well. Jocko was friendly, polite, respectful. But afterwards, he dropped Bourgin off first.

The next day, Jocko called Eleanore at her brother's house and asked her out to dinner. He seemed very jolly to her that night, fun and full of energy. She was surprised that he knew so much about movies and actors. His favorite actress was Wendy Hiller, an aristocratic Englishwoman who had been a smash in *Pygmalion* in 1938. And he was so smart, it was as if he remembered every book he'd ever read. He could quote from things he'd read in high school.

After he took her home, he stayed to talk business with Dean and she went upstairs, then snuck back out onto the staircase, crouching down and peering through the rails to overhear what they were saying. He was so charmed he still remembered it decades later—it was so sweet, so girlish.

At first, they only saw each other on weekends, going to parties or to dinner or to nightclubs where they listened to Anton Karas play the zither. *The Third Man* was shooting in Vienna that fall and sometimes they stood on the street and watched them shoot scenes. She talked about growing up in Michigan, where her father was the city editor of the *Detroit News* until he quit to go into advertising, where she joined the Brownies and got Bs and Cs at Edwin Denby High School in East Lansing. She told him about studying at Columbia University in New York, where she lived in a cold-water flat on the Lower East Side and took showers at John Steinbeck's old apartment, which had a broken egg painted on the kitchen floor that was so realistic you stepped around it. She probably didn't mention her flirtation with the Progressive Party, or Dean and Nancy living in sin before they got married, or the paper she just finished arguing for a "synthesis between capitalism and communism under some form of world government."

One weekend he came up to Linz and helped her take the kids from the Youth Program skiing, which was a good thing because she had new boots and got terrible blisters and he ended up practically carrying her down the mountain. She was impressed with his excellent German, which helped to keep the kids in line. And since it was a Sunday and neither of them cared to go to church, they began to talk about religion. He told her he'd given it up and she told him that after her father died—it happened suddenly, on a golf course, when she was just eleven—they went to live with a Catholic grandmother so pious she went to church every day and disowned her daughter for marrying a Lutheran. That experience turned her away from religion forever. So they were both freethinkers, another point of connection.

Soon the drive out of Linz started getting scary. Once a soldier stepped in front of her car with his hand out and more soldiers appeared with guns. They said they were looking for a deserter and ordered her out of the car, but she refused. She'd been warned that the Soviet guards would say anything to get you out of the car and then they'd snatch you and say they'd never seen you. Finally they checked her trunk and let her go. Another time she went over a bump and the oil pan came off and she ended up pushing the Topalino with her foot through the turns of the Serpentina Weg—all downhill and twisting like a snake. She almost lost the car.

So Jocko and Dean decided she'd be safer if she started taking the

Mozart Express. But the hard wooden train seats were very uncomfortable and one time a man tried to get her to carry a package for him, pleading and begging and telling her "it's for Mama in Vienna." After that Jocko and Dean decided she better not do that either. So she went back one more time in Dean's new convertible, a bright red thing that looked like a torpedo with its flashing fins and gleaming fenders, and when she got to the checkpoint at the Enns Bridge, the Russians wouldn't let her go. They wanted to get in the car. They wanted to feel the upholstery. She kept fighting them off and arguing in bad German and then the American soldiers on the other side of the bridge could see she was being held up and started to approach the no man's line at the middle of the bridge. And finally the Russians gave in and told her to go.

By then, Dean and Jocko were waiting at the checkpoint, panicked and angry. "You're not going back again," they said.

So she said it was time to go home.

"Or you could marry me and stay here," Jocko said.

That surprised her. They had only known each other six or seven weeks, had kissed a few times and nothing more. Later she would think it was mostly that he'd been through the war and was ready to get married. But she said yes. Not long after, he came over to Dean's house and said he wanted her to go with him to pick out a ring. They went to a jeweler named Heldvein's, a famous place right across the street from the statue commemorating the Black Death. Jocko told old man Heldvein he wanted a big diamond and Heldvein brought out a glass case with loose cut diamonds on velvet. Jocko pointed to a beautiful three-carat round white diamond and Heldvein held it up with tweezers. Yes, that's the one. "I will set it myself," said Heldvein. Only white platinum would do.

At a party in the Press Club, Jocko announced the engagement. The teasing started immediately. "If you're going to marry Jocko," the men told her, "you have to learn how to drink martinis, smoke, and play poker!"

They got permission to get married at the Rathaus, the Vienna City Hall, splurging on the hundred-dollar wedding with a red carpet on the beautiful marble staircase. Jocko asked her to fill out extensive forms about her life and education, saying he needed it because he worked for the embassy. At this time she still thought he was "second secretary."

At some point, I'm not sure when, they went to the lake country for a romantic weekend. Although it's my impression that my mother was not technically a virgin, all she would say for the record was: "I was innocent—he wasn't."

They got married on February 25, 1950, only ten weeks after they first met. He was thirty-six, she was twenty-five. In later years neither one of them ever mentioned the actual ceremony, never said whether they were excited or frightened or astonished to find themselves there. I know that Dean was the translator and Messing was the best man and when they came down the red carpet, the mad Serb who ran the Balkan Grill handed Jocko a baby white pig, telling him it was a Hungarian symbol of fertility. So Jocko carried the pig to the Hotel Bristol and held it in his arms in the reception line, and then they danced and went off on their honeymoon with tin cans tied to the car and a Just Married sign, driving right through the Soviet zone and stopping just on the other side in Gmünten for the night. They drove through terrible snowstorms in the Brenner Pass to Lago di Garda and Sorrento and Rome and Florence and, after stopping to visit the home of Catullus, ended up in Capri, which my mother would always remember as "cliffs and cliffs and cliffs."

16

THE SQUIRREL CAGE

After the honeymoon, they returned to the house on Fuerfanggasse. Outside there was a strawberry patch and a small vineyard. They had a piano and Jocko used to play for her, waltzes mostly. They had a study with armchairs and bookshelves. They even had a sleepy old Irish setter. She shopped at the PX and a housekeeper named Friedl ran the house and Fritz ruled the kitchen, particular about everything from the strudel to the crease on the linen tablecloths. There was a bodyguard who slept at the foot of the stairs every night, but at first she didn't think much of it.

At times, it was a bit difficult for her to adjust. She was only twenty-five, and there were so many new rules. An embassy wife does not leave her house without her gloves or her hat. When arriving at a new post or a first post, she makes calls on everyone higher in rank, presenting a card with her address written in pencil in the lower right hand corner. People from lower ranks call on her, presenting their cards.

One day she was up a ladder in shorts and an old shirt, painting the walls, and in walked Ambassador Erhardt's wife in her formal hat and white gloves. Eleanore blushed. "I'm so sorry I'm not dressed," she said. But Mrs. Erhardt was very gracious. "Don't be silly. You didn't know I was coming."

She got a copy of the *Vogue Book of Etiquette* and consulted it frequently.

But when they went to a luncheon at the ambassador's house, she saw everybody dipping spears of white asparagus into a sauce with their fingers. Their fingers! It was enough to make you dizzy.

As the weeks went on, she began to learn a few things about the man she had married:

He always kept a copy of the *Meditations* of Marcus Aurelius next to his bed.

He took sleeping pills every night and drank fifteen or twenty cups of coffee a day from his secretary's big silver coffeepot.

He always wore Borsalino hats, except when he wore a beret.

He was squeamish about blood. She cut her knee one day and asked him to wash it out and he couldn't. He admitted that one day as a student in Berlin, he got a shot, walked three blocks, then passed out in the street.

He didn't want her to become one of those bridge-playing women who spent all their time at card parties with other embassy wives, something she agreed with completely. She wanted to get to know Vienna and explore a foreign culture and *learn* something.

And he loved to buy her clothes. When she realized she didn't have much of a wardrobe, he became fascinated with the whole process, going to fashion shows and even meeting designers when they came in from Paris. He bought her a beautiful Dior gown and a black afternoon dress by Jacques Fath and a full-length Balenciaga mink coat that irritated the wife of the DCM—the Deputy Chief of Mission—because she didn't have one. And they were invited to dance the first dance at one of the balls that were held in a huge ballroom with tiers of seats and a giant dance floor. It was a linx waltz, moving to the left instead of to the right, so Jocko and all the other men took dance lessons. On the big night, he wore black tie and she wore the Dior.

Perhaps he was thinking of Wendy Hiller in *Pygmalion*.

He also gave her a collection of the writings of Jonathan Swift, and wanted to discuss it. Then there was another book and another book and finally she put her foot down and said she was going to read what she wanted to read and she wasn't going to submit to a deposition either.

He wasn't too happy about that.

Another little marital power struggle developed when Jocko decided to get in better shape and bought a gym set with handle bars and rings. She asked him to put it down at the end of the yard where it would be less unsightly, but he put it right next to the terrace, exactly where he wanted it. He had that stubborn streak.

Sometimes puzzling things happened, like the time the DCM's wife called up concerned because she couldn't find Jocko on the payroll. So Eleanore called him and he said not to worry about it, it was just a mistake. Sometimes he'd send her upstairs and tell her not to open the door until his guest left, or ask her to leave so he could brief the reporters. One night he came home in a dark stew and said that a policeman stopped him and argued with him and he got so mad he punched the policeman in the face. Sometimes he had to go to Salzburg and Munich, taking off in a little airplane from a narrow airstrip on the bank of the Danube—a hair-raising experience reserved for urgent government work. But he never told her why, or what he actually did for a living.

And then one night she opened the drawer of the desk in the study and saw a gun. "What the heck is this doing here?"

"I have that for protection," he said.

So he decided to show her how to handle a gun. "The first thing you do with a gun is take all the bullets out. Then you make sure there's no bullet in the barrel. You do that by going out and shooting it into the ground."

He pulled the trigger. A bullet hit the armchair two inches from Eleanore's shoulder.

That upset him terribly. He told her about what happened to his brother and stayed nervous for a long time. But finally he calmed down and she tried to ease the mood more by teasing him. "We haven't been married very long. If that bullet had hit me, you'd have been in big trouble."

The story got around, turning into a bit of black humor.

Finally he told her what he really did, although by that time she had pretty much figured it out. He said she couldn't talk about it ever, she always had to say he was the second secretary at the embassy and nothing more. If they persisted with questions, deflect them with a story. And she

couldn't expect him to come home and talk about what he did that day—
he was not going to be that kind of husband. "I took an oath of silence," he
said.

It turned out there wasn't much to do in Vienna. Jocko taught her how
to play poker and they had a regular game with some of the other couples.
On Sunday mornings when most people were in church, they would play
softball with the reporters. She taught him to play golf on the little course
the Viennese had built inside the race track, where you had to wait for the
horses to go by before you could drive. She did some volunteer work, collecting food and coal for crippled children, taking them to movies on Army
buses. Sometimes they would have dinner at the Drei Husaren or go to the
Augustiner Keller, a huge place with wine and dancing. And there were
parties at the Bristol, including one wild reception when Jocko got plas-
tered on a lethal punch made out of champagne and cognac. When they
got home she realized she'd forgotten her hat and he wanted to go back for
it, but he was so drunk she ended up locking the door. He just climbed out
the window, coming back an hour later with the hat and a bleary grin.

It didn't seem unusual. Everyone drank like that in those days.

And the tension level was higher than ever. They'd been married just
under a year when Howard Hunt got a call to identify a body at the Vienna
morgue. Later to become famous for his participation in the Watergate
burglary, Hunt was then a civilian working for an economic aid agency. As
he told the story later, he arrived at the morgue to find the body of a young
colleague, his skull beaten soft. Although the man was not a spy, he had
been in contact with a Navy captain named Eugene Karpe who had been
in contact with a man named Robert Vogeler, who had recently been sen-
tenced to a long prison term in Hungary after being accused of plotting to
sabotage the state telephone company. Later that winter Karpe turned up
dead in a train tunnel near Salzburg, his body spectacularly mangled and
smashed.

Immediately, the embassy reminded everyone of the evacuation plan: if
the Soviet troops came across the lines, the women were to put on dirndls
and the men leiderhosen and walk out of Vienna through the woods.
Eleanore and her friends scoffed at both the plan and the fear, but others
were not so brave. The provost officer's wife got so scared she took her
children back to the United States, teaching another important lesson

about the duties of embassy wives. From that moment on, there was a cloud over the provost officer. Nobody was surprised when he went home to join her.

At this point, my father got tangled up with a couple of prominent war criminals. The first was a former SS man named Wilhelm Hoettl. In 1948, a cable from my father's office—although the identifying details are marked out by the CIA censors, I'm pretty sure he wrote it—passes a tip that Hoettl was trying to develop contacts in Switzerland and Romania. Another in 1949 cites a report that he's hidden a nest egg of stolen Jewish property. By 1952, Hoettl seems to have exhausted everyone's patience. "Any action taken to frustrate, impede, or discredit Hoettl's operations will be appreciated here," one cable says. "From the point of view of the Austrian Station, support to Dr. Wilhelm Hoettl is a Bad Thing."

The biggest gripe seems to be Hoettl's habit of picking up useless information and "dizzying" it to as many buyers as he can. "Hoettl, like a couple of dozen other types in Austria, can turn out a sizeable volume of such information without interrupting his regular pattern of coffeehouse conversations."

The next cable in the file is definitely written by my father—for once, the CIA censor forgot to wield his marker. This one is about two CIA officers who went to inspect some new office space in Linz, where they met "an intelligence operator of the 'breathless' school." All excited, this breathless officer said he had obtained some very valuable information from an agent of a German intelligence service who wouldn't give his name. The CIA men agreed to have a look but my father was skeptical, especially when he found out that the breathless agent had already paid for the information and now expected to be reimbursed. And sure enough, it turned out that his suspicions were correct and the unidentifed German agent was "the master fabricator Hoettl," the information the usual pack of lies.

Furious, my father endorsed a plan to "burn" Hoettl. That plan went into effect seven months later when the CIC arrested Hoettl in connection with yet another espionage ring, this time a pro-Soviet operation run by two American citizens named Kurt Ponger and Otto Verber. At his home, they found a pistol and a false passport. After putting him in a chilly cell with no windows and interrogating him with a lie detector, they decided he was probably innocent this time, but held him for two more days and leaked a false story to the press. My father's revenge appeared in the *Washington Post* on April 7, 1953:

> U.S. Army authorities said they had picked up Wilhelm Hoettl, a former Nazi SS (Elite Guard) officer who engaged in espionage in the Balkans for the Germans during World War II. The Army said he had connections with Kurt L. Ponger and Otto Verber, naturalized Americans arrested in Vienna and charged with giving U.S. military secrets to Yuri W. V. Novikov, second secretary of the Soviet Embassy in Washington.

Reading this file, I'm amused to see my father manipulating the press, impressed by his command of this slippery world. It's good to see him showing no patience for the bullyboy types, which is consistent with the CIA's overall record in those days despite headlines to the contrary. Military intelligence was much worse.

But as I read on, things get murky. The second set of cables are about Otto von Bolschwing, the same man his team sent packing in 1947. As an assistant to Adolf Eichmann, von Bolschwing developed plans for terrorizing the Jews out of Austria. As Heinrich Himmler's representative to Romania, he supported the ultra-Fascist Iron Guard during a bloody pogrom of the Jews of Bucharest. But he wormed his way into a job working for the German branch of the Org in 1947 and managed to hold onto it after the CIA started funding the Org in 1949. Recent scholarship suggests that some intrigue was involved. In a 2004 study of freshly declassified documents, a historian named Timothy Naftali determined that the German CIA station even went to some lengths to clean up his file, possibly out of the fear of being embarrassed. It was a terrible mistake. "The moment the CIA acted to whitewash von Bolschwing's past," Naftali said, "this Nazi war criminal gained enormous leverage over the U.S. Government."

A year later, trying to get rid of von Bolschwing, the German CIA station transferred him to my father's Salzburg office. He was assigned to develop contacts with his old friends from the Iron Guard.

Bill Hood took over as my father's deputy chief soon afterwards. He told me that von Bolschwing was a con man who never did a damn thing. "I didn't like him and your father didn't like him." But the record shows that six months later, von Bolschwing was promoted. A year after that, just around the time my father was leaving Austria, the Agency reversed itself and shut down von Bolschwing's network. Then something even stranger happened. "Instead of merely firing him," Naftali said, "the CIA station did something unimaginable. It chose to reward this incompetent by helping him achieve his long-term goal; the CIA decided to help him become a U.S. citizen."

The shreds of censored cables don't tell me what part my father took in this, if any. But Hood said he thought the strings were pulled in Germany or from headquarters and the Austrian station didn't have anything to do with it, and there's one hint that the decision did come from the top, a memo from a CIA chief in Washington to the immigration bureau that says von Bolschwing joined the Nazi Party reluctantly and went to prison for pro-Jewish sympathies and even fought for the Austrian resistance, all things the Agency knew to be lies. Beyond that, even though von Bolschwing was discovered living in the United States in the 1970s and all of this became a national scandal, we still don't know why the CIA promoted and sponsored him. Too much is still classified.

But it looks like that corny old spy movie line applies—von Bolschwing knew too much.

When spring came, Jocko took Eleanore and some friends on a pleasure tour of the Voralberg, a mountain range of dramatic peaks and glaciers at the western tip of Austria. The photographs of this trip have the glamour of a true suspended moment. Eleanore looked particularly stunning with her lush lips and Italian sunglasses, and one of the men even managed to snap a few scandalous pictures as she changed into her bathing suit. In June they went to the Bavarian Alps to see the Passion Play at Oberammergau. On the way back, they stopped at the officer's club just outside of Linz to have lunch and heard the news. North Korea had invaded South Korea. "We have to go right now!" Jocko said. "The Soviets will close the checkpoints! Hurry."

They rushed to the car and took off with screeching tires, hitting the bumps and dips of the road so hard that the car left the ground for agonizingly long stretches. Eleanore kept thinking they would just keep on going, a flying car.

That last checkpoint was agony, but they got through just in time.

Two days later, Jean Nater got a tip that a Soviet commander at the Zistersdorf oilfields had ordered the large oil storage tanks cleaned out and readied for immediate use. Maybe they were planning to stockpile oil to support a strike at Vienna?

Jocko took the news to G-2.

After all that, the trains were too dangerous and they couldn't risk driving. So they were trapped in the squirrel cage. At work, Jocko and his team concentrated on developing a stay-behind program, recruiting Austrian radio operators and sealing transmitters in gasoline cans, burying them at selected spots just inside the Soviet lines through the Vienna Woods. Nater and a few others volunteered to stay behind and walk out after a Soviet invasion, and Jocko authorized the proper gear: Austrian mountain boots made to measure, rucksacks and Austrian clothing, dirndls for the women. Through the most trusted local contacts, they got Austrian identity cards, Nater using the name of a dead friend whose family agreed to go along with the ruse and even to take them in if the Soviets came looking for them. Jocko ordered him to draw up an escape plan—Project Lace, they called it—and Nater and his wife began a period of "lovely long walks with the dog in the afternoon and the early evening, noting landmarks and walking paths." They also made plans to train and equip a core of resistance fighters, burying weapons in the forests. Every so often a gardener or game warden uncovered a cache, but the Austrian government never got very excited. And headquarters pressured them to send border-crossers into Hungary and Czechoslovakia to give early warning of a Soviet attack, so there was a constant search for viable refugees who might take four or five thousand dollars to go back.

In October, the Soviets tried to foment a coup by trucking Austrians in from the Soviet zone with instructions to riot. Soon the streets teemed with men pushing and shouting and the air was jittery with danger. At one point, the protesters pushed so hard that the Viennese police almost lost control and the Austrian authorities became so alarmed they asked the Americans to send in troops. In the Allied camp, urgent discussions followed. What if the Soviet ringers resisted? What if American troops started fighting on the streets? Would the Soviets use that as a pretext to send in their own troops?

They turned to Jocko. Working with his friends and contacts in the

Austrian Interior Ministry, he came up with a plan to send the construction and transport workers into the streets armed with axe handles, so many axe handles they had to bring them in on trucks—an operation expensed to Loch Campbell's OPC budget. So the workers took on the worker's state and won, a classic example of using the CIA to preserve America's plausible deniability and hold off the risk of hot military entanglements.

Around that time, one of the Austrian station's most valuable agents was snatched. I think he was the man my father described in a letter as a "courageous, daring" police major, a close friend who disappeared on his way through the Soviet zone to Linz. As Bill Hood told the story, he was an impressive and sympathetic man who had become one of the treasures of the station, so valuable that they used him in too many efforts and let him know too much inside information. In his wake, major changes had to be made to limit the damage. Agents were retired, some sent to the United States. Networks were shut down. For the station, it was a painful education in "compartmentalization."

Still, the day came when someone in an elevator made a remark about my father's job at the CIA, and the elevator operator was a secret Soviet agent. From that moment on, his name was in KGB files.

Trapped in the squirrel cage, they played poker and golf, tennis and charades, got married and divorced. A top embassy officer got caught with a male prostitute and quietly left the service. Uncle Dean started a secret affair with a glamorous Czechoslovakian refugee named Renata, splitting up with Nancy. Nancy took up with Ted Kaghan, the embassy press officer. Kaghan cancelled his wedding plans.

Eleanore got pregnant.

In the summer of 1951, they were all shocked by Senator Joe McCarthy's speech attacking General George Marshall and the Marshall Plan. Rebuilding Europe was Communist appeasement? How could he say that? What did these right-wing bastards want to do, just abandon the whole continent and then come back in thirty years and fight another war?

And Eisenhower didn't say a word.

Except for one die-hard Republican, the whole embassy was pulling for Adlai Stevenson. They didn't want a general as president. The war was over. But Stevenson lost and then all the partisan ugliness arrived on their doorstep in the form of Roy Cohn and G. David Schine, a pair of ambitious young McCarthy disciples come to hunt for Communists in the American embassies of Europe. When they got to Vienna, they checked into the Bristol and held a press conference, asking the assembled reporters—Si Bourgin was there—to finger the reds in the embassy. Somehow they fixed on Kaghan, who had written an anti-fascist play in college called *Hello, Franco* and also signed a petition put out by some communist labor group. So they tried to squeeze him for names and Kaghan got his back up and called them a couple of "junketeering gumshoes," a phrase that ran in newspapers all around the world and stuck to them for years. They got their revenge by summoning Kaghan back to Washington to answer questions in front of McCarthy's committee, a summons he was afraid to resist because Nancy actually *was* a Communist. Even back when she and Dean were living in sin in New York, it turned out, she was sneaking off to a party meeting every Thursday night. So Kaghan was fired and his career ruined.

From Vienna, Jocko and Eleanore watched in disgust. Jocko had always been very fond of Ted. Ted was with them at the Voralberg, the one who took the sexy pictures of Eleanore getting into her bathing suit. And he was almost part of the family now, stepfather to Dean's first child.

Then they were distracted again, this time by the arrival of Jennifer Lynn Richardson, who made her father so proud that every time he brought guests home, no matter how late at night, he'd lead them all upstairs to wake her up and show her off.

17

*

JOEL ROBERTS

Here the curtain parts again, courtesy of an unusual book called *Mole*.

The year was 1952. At the CIA headquarters in Washington, Dick Helms had a last meeting with a dashing young spy named Bill Hood. "Your job is to recruit Russians," he said. "Until we've done that, we've failed."

Only twenty-five years old, Hood started his professional life as a seventeen-year-old boy reporter for the *Portland Press Herald*. When the war started, he enlisted and joined the OSS counter-espionage team in London, recruiting double agents and putting out false invasion plans to confuse the Germans, then joined the CIA afterward and served with Helms in Germany. Now the new chief of operations for Austria, Hood arrived in Vienna and reported to Jocko Richardson. By reputation, Hood knew Richardson as a respected man, well liked within the agency, but there was always the possibility that he'd get defensive and treat him like a snot-nosed kid from Washington. So he was diplomatic, saying that the fault was really at headquarters for failing to make its changing intelligence needs clear. Although Vienna was crawling with recruitable Russians, for example, they still hadn't sent out a single full-time Russian-speaking case officer. But now that had to change. There would be no more reports on the Czech Communist Party and Soviet wheat shortages. From now on, the first priority and the second priority and the third priority would all be the same: penetrating the Soviet Union.

Fortunately, Richardson wasn't defensive. On the contrary, he gave him the keys to the shop and even piled on the responsibilities. So Hood plunged into action, his first goal to get mug shots of every Soviet staff member working at the Hotel Imperial. He ended up parking a car across the street from the hotel with a camera hidden inside a stuffed animal on the dashboard. Every time a Soviet officer came out of the building, the stuffed animal blinked.

But the real job was a daunting one. In the seven years since the end of World War II, the combined forces of all the intelligence agencies of all the Western nations hadn't managed to recruit a single significant Russian official. First-hand information was so hard to come by that every Soviet soldier who deserted got airlifted to Germany for an immediate debriefing at a special defector center. Once, when an actual Soviet spy tried to defect, the British turned him over to a senior agent named Kim Philby and the defector was promptly kidnapped, drugged and shipped back to Moscow on a stretcher. Later they discovered that Philby was the most notorious Soviet spy of all.

Their breakthrough came by accident. One day the wife of a young American embassy official found a letter addressed to "An den Americanisher Hockkommissaer, Bolzmanngasse, Vienna XVIII." She gave it to her husband, a junior American diplomat named Horace "Tully" Torbert. The letter had no stamp and the grammar and the spelling was wrong, all signs that something interesting might be inside. But Torbert was in a hurry and stuffed it into his pocket unread. It was late afternoon before he realized there was another envelope inside. Inside that envelope, he found a letter written in the Cyrillic alphabet.

Many years later, I interviewed Torbert in the Metropolitan Club, an elite Washington private institution two blocks from the White House. By that time he was the distinguished Mr. Ambassador, long returned and living in a retirement community. But he still remembered what he did with that letter. Of all the spies crawling around Vienna, he said, my father was his favorite, a man he respected as a scholar with deep principles. So he took the letter straight to him.

This is where Bill Hood picked up the story in *Mole*, the first inside story of a CIA operation and still one of the best. The letter was quickly translated and he met my father in his office to study it line by line:

> I am a Russian officer attached to the Soviet Group of Forces Headquarters in Baden bei Wien. If you are interested in buying a copy of the new table of organization for a Soviet armored division, meet me on the corner of Dorotheergasse and Stallburgasse at 8:30 p.m., Nov. 12. If you are not there I will return at the same time on November 13. The price is 3,000 Austrian shillings.

To placate the CIA censors, Hood called my father "Joel Roberts." But he said that everything else in the book is true and for me, the bell of recognition went off with the first anxious question Joel Roberts asked: "Have you ever heard of a bona fide approach as bald as this one?"

Of *course* he starts out worrying! And Hood said it was also typical behavior for Roberts to defer to him. Never having served in CIA headquarters and humble by nature, he tended to assume people from the home office knew far more than he did.

Next, he worried about the meeting point. The corner of Dorotheergasse and Stallburgasse was quiet and exposed, and deep in the Innere Stadt, a place where the KGB operated freely and kidnappings were common. The only upside was visibility. "At least we should be able to see if he's alone."

City maps confirmed the thumbnail analysis and Hood tipped his hat to the boss: "After six years in Austria, Roberts knew every alley in Vienna's Innere Stadt."

But Roberts had only begun to worry. The only Russian speaker they had available was a refugee whose native Russian accent made other refugees nervous. And since he wasn't an American citizen, the Soviets could easily snatch him and get away with it.

For the table organization of a single Soviet armored division, it wasn't worth the risk. But if they could recruit the man who got his hands on it and set him to work as a double agent, that would be a prize indeed. So they sent a case officer and an operations team to do all the spy stuff you see in movies, one man in a car with a secretary to handle the walkie-talkie, a meeting on the sidewalk and a quick drive to a safe house where a camera clicks away in the other room. They need to document everything in case they have to blackmail the man into a second meeting.

By the time I interviewed Hood years later, he was a craggy and sophisticated man in his seventies wearing an ascot and a neat little white moustache, the author of five intelligent books on the CIA. On his wall, there was a framed poster of Miles Davis and Charlie Parker. He downplayed the spy-movie stuff, saying they didn't even take weapons because my father didn't approve of them and anyway, they always felt it was better to talk their way out of trouble. There's a personality change when you carry a gun, he said. You start to get brave. The real story was what happened the next morning, with the overnight cables and the secretary who brought in coffee and donuts on a silver tray. The case officer barged into the room, too excited to wait on protocol. "The guy was there, right on schedule. Alex took him right to the safe house. He didn't bat a goddamn eye."

My father reached back and turned on the radio to mask the conversation, a bit of extra caution above the usual security sweep to check for bugs. Then the case officer told them the man brought photocopied documents that seemed geniune, though not too exciting. He refused to identify himself but his clothes and accent were clear signs of a native Russian. Clearly, it was an exploratory meeting, a getting-to-know-you moment.

When they talked it out from every angle, my father sat down and wrote the first "blue bottle" cable of his career, this being the Agency term for messages so secret they can only be decoded by the chief of the communications staff at the moment they arrive. After detailing the approach and the meeting, he drew a tentative conclusion. The Russian's civilian clothes and his access to a photocopy machine suggested either a diplomat or an intelligence officer, though he was leaning toward intelligence officer. The man seemed to need money and didn't seem interested in defecting, both extremely good signs. So while it was possible that he was trying to con them in some way, perhaps as a sting operation, and they would continue to be very cautious every step of the way, could they *please* have a seasoned Russian-speaking case officer as soon as possible?

Twenty-four hours later, he got his answer. From this moment on, headquarters said, everything else in the station came second. The Russian-speaking case officer would arrive on the next plane.

Studying the cable, Hood and my father noticed one thing missing. No desk officer would have sent a cable without including the traditional Agency signoff, "proceed with caution." Which meant that Helms must have written it himself.

From that moment on, my father imposed strict compartmentalization. Absolute secrecy had to be practiced. No one without a need to know would get even the slightest hint of the new operation. That meant cover stories and coded cables and constant supervision of the office staff to keep them from blurting something out or leaving some detail exposed, soon to grow into an extremely time-consuming obsession.

A day later, the Russian-speaking officer arrived from Washington, along with a backup man who specialized in Soviet intelligence. They set up a second meeting at the same safe house and this time things went quite smoothly. The Russian spy seemed to relax. He said that his name is Pyotr Semyonovich Popov and he was a case officer for the GRU, Soviet military intelligence. He described the structure of his office and named the men who ran it. While the Russian-speaking case officer struck up a friendship with him, a hidden camera snapped his picture through a pinhole.

At the end of the meeting, my father's men gave Popov his first payment, about a hundred dollars.

When they got back with their report, my father gave in to his excitement. "Tell me something more about our friend? Who is he and what's his background?"

They didn't know much. Popov was about thirty years old, not long out of school. He had been in Vienna less than a year.

They checked the names he gave them against the names on the charts in their operations notebooks and discovered that they'd done a decent job of spying. A third of the people Popov mentioned, they already had in the files. "Why is he doing this? He must have known he didn't have to give us his name. He could have strung us along for weeks with an alias."

Gambling debts, a girlfriend, a black market deal gone bad? They still didn't know. Popov said he was in some kind of jam but wouldn't say what.

"What about his family? Is his wife here with him?"

They didn't know that either.

More meetings followed. Over the next few weeks, Popov gave them a road map to the GRU's operation in Vienna, seventy officers in fourteen different sections under the command of the "counselor" to the Soviet High Commission for Austria. The head of the Anglo-American section was, as they suspected, the same Colonel Sitnikov who drank so much at all those diplomatic receptions. He told them about the warehouse where they keep radios, truth serums and disguises. After the safe house meetings they had to transcribe the tapes and mine the transcripts for every useful detail and then arrange all that information into some kind of useful form, doing the filing and cross-indexing and all the other boring stuff that makes raw intelligence useful. They also got to know him better and better. He came from a peasant background and drank too much and took too many risks and hid, deep in his heart, an abiding hatred for all the masters and rulers with whips and shiny boots. Before long they were all convinced—if Popov was trying to infiltrate them for some kind of elaborate double-agent scheme, he wouldn't have told them so much. He would have fed them small details, morsel by morsel.

But CIA headquarters balked. Why wasn't Popov asking for anything substantial? Why was he risking his life for a few hundred dollars? Was he unstable?

Finally Popov admitted that he had a girlfriend, a bad sign. A spy with a weakness is a vulnerable spy. So Hood ran a police check on her name and sure enough, there was a police report about a Captain Popov showing up drunk at her door. There was even a CIA case officer assigned to look into it, a process guaranteed to attract curious eyes.

When my father found out, he got so upset he barked over the noise of the radio. "This is something we can damn well do without."

Gingerly, Hood handed him the police report.

Even worse, it turned out, Popov had put his mistress on his payroll and passed her off as a source. That's something spies try to put over on their bosses, and when the bosses find out they ship them straight to Siberia. For an hour, they argued over what to do. Finally my father wrote a cable outlining the problem and his conclusion, that they should work to deepen their relationship with Popov and save any ultimatums about the mistress for later. Otherwise they might scare him away.

After he sent it, he stewed in anxiety. What kind of a plan was hoping for the best? But the CIA agreed, and over the next few months, the operations team met with Popov whenever they could. My father met him once, just to get a feel for him. And it really seemed as if Popov was the dream double agent, the spy who didn't want to defect and didn't want to get rich, who would stay in place and provide fresh intelligence for reasons of his own. In cable after cable, my father pushed this view.

Then the CIA merged the OPC and the OSO, putting my father in charge of the covert action arm of the station. In the wake of the reorganization, headquarters gives my father new orders: instruct Popov to form "a small, tightly knit resistance group of like-thinking comrades."

He was appalled. Risk a spy this valuable on covert action? Trying to contain his indignation, he wrote what Hood described as "a polite cable" suggesting that headquarters reconsider.

The response was swift and frosty: his objections had been noted at "a high level" and he should proceed as directed.

He obeyed. At his instructions, his men asked Popov to form a resistance group. But Popov immediately dismissed the idea, giving them a grim lecture on Soviet reality. In his entire life, he said, he'd never met a single person who so much as hinted at disaffection. Even his friends who seemed like the kind of people who might dissent could be spies for the KGB, so he was afraid to say anything to them. If he tried to feel them out, they might think *he* was spying on *them* and report him just to cover themselves. He couldn't even talk to his wife. Ironically, the only people he could tell his secrets to were the men of the CIA.

My father put all this in a report, and the CIA bosses in Washington finally agreed to table the idea. For years afterwards, the spies of the Vienna station used "a small, tightly knit resistance group" as in-house code for "another wildly unrealistic idea from headquarters."

Then headquarters asked if Popov could get a copy of the 1947 Soviet Army field regulations, the basic Soviet manual of army tactics and organization. For five years, this had been one of the dearest items on the Pentagon wish-list. But when the case officer asked him, Popov looked puzzled. "I might be able to bring in one of the 1951 manuals," he said, "but I don't know where I could find the old one."

This was the first time they heard there *was* a new field manual, launching the Popov team into a series of unprecedented intelligence coups that ended up saving the Pentagon more than a half billion dollars and even helped to unseat the head of the KGB, one of the most fruitful spy operations in American history. But here my father's piece of the story ended. Rewarded with a promotion to chief of the southeastern Europe division in Washington, D.C., he packed up and said goodbye. After ten years in Europe, he was heading home.

A few years later, Popov was summoned home too. By this time, his CIA handlers had become close to him and respected him as a decent fellow. With the money he earned spying, all he did was support his mistress and buy a new cow for his family back home. Hood's unflinching description of his fate gives a hint of their feelings:

> Popov would have been held incommunicado in the cellars of the Lubyanka prison, adjoining the rear of the KGB headquarters at Two Dzerzhinksy Square. The KGB's first objective in a forced interrogation—an interrogation in which the victim may be savaged at the whim of his questioners—is to reduce the prisoner to the lowest possible mental and physical condition. At the outset, Popov would have been beaten with professional skill, badly enough to convince him that resistance was impossible and that he was powerless in the hands of his jailers . . . to increase pressure on Popov he may have been permitted to see his wife. Perhaps the abused and broken woman was allowed to plead with him to cooperate with her tormentors. The impact of Popov's appearance on his wife could only have added to her despair.

Reading this, I recognize the tone of relentless and almost ruthless precision. I wonder if this is the sort of thing my father was thinking about, all those years when I used to go into his room and find him awake, drinking and brooding and burning a chain smoker's rosary into his bedspread.

John H. Richardson Sr. in his high school graduation photo, 1931.

With his younger brother, Leonard, in the late 1920s. Leonard died under mysterious circumstances in 1938.

Stuck stateside, waiting for his war to begin.

Richardson and the other soldiers in his CIC unit supplemented their rations with fresh pheasant, dandelion salads, and liberated wine, Caserta, Italy, 1944.

Portrait of a cold-war station chief, Vienna in the late 1940s.

In Venice, celebrating the war's end in 1945. A rare moment of complete relaxation.

On a fishing trip near the Albanian border with Gordon Mason (*right*) and Peter Kormelis, 1956.

With a brand-new master's from Columbia, Eleanore Koch arrived in partitioned Vienna in 1949.

My parents at their wedding, Vienna, February 25, 1950. The pig was a traditional Hungarian fertility symbol, given to my father by the mad Serb who ran the Balkan Grill.

Our first Christmas in
Athens, December 12, 1955.

He became so famous as a fisherman
in Greece that the queen insisted he
take her with him. His favorite spot
was right near the Albanian border.

In Athens ceremony was vital, and my parents frequently
consulted with the embassy "protocol officer."

A rare domestic moment, Forbes Park, Manila, 1959.

My mother dancing with the head of Philippine intelligence, a frequent guest at our parties.

Father and son, 1955.

In the backyard
of our house in
the Philippines,
1961.

Inspecting Special
Forces troops in
the mountains of
Vietnam, 1962.

Getting to know General
Ton That Dinh at Cap
St. Jacques, Saigon, 1962.

In McLean, Virginia, one street down from CIA headquarters, he built his house upon the hill. Two years later he sold it so he could get back into action.

Not the warmest Christmas: While my father fought one of the last stands of the cold war in Korea, I was moving in a different direction, Seoul, 1971.

In Korea the drinking parties were ritualistic and endless, with geishas always waiting to refill your glass.

Bitter at the purges of the CIA's old guard that began in the early 1970s, he delayed his retirement ceremony until he could get his send-off from his old friend Bill Colby, Washington, DC.

Reunited again in Mexico, 1975.

In his seventies, finally relaxing with his first grandchild, Christmas, 1987.

18

•

VIRGINIA, 1953

T he Messings had a television. After we got off the ship in New York, that was the first amazing thing. My father would stay up watching until the last show ended, and never lost his bemused fascination with American progress, like someone watching a child who had just done a very impressive trick.

Arriving in Washington, my parents bought a used Mercury sedan and went in search of houses to rent. From a friend at the CIA, they heard that Lyndon Johnson was renting his house for the summer, so they went to the Senate office building for an interview. My mother remembered Johnson as "huge behind his desk" and very pleased to hear they'd brought Friedl from Austria to take care of the house and the baby. The rooms were full of pictures of Johnson. At the top of the stairs, there was a shot of a small Texas house with a caption: Childhood Home of the Future President of the United States of America.

When the summer was over, they found a lovely house in Bethesda with a tennis court and two huge oak trees in the yard. But when my mother tried to get a telephone and put down her usual cover story, that her husband worked at the State Department, the phone company called the State Department personel office to confirm his employment and found out he wasn't on the books. Someone from the CIA had to call and explain.

My father went to work in the CIA's temporary Quonset hut offices near the Washington Monument. As director of the "Balkan" division, he was responsible for Albania, Yugoslavia, Greece, Bulgaria and Romania, plus a few outposts in Frankfurt, Paris and Rome. Or so I learned from Howard Hunt, one of his deputies at the time.

At home, my mother bought most of the furniture that became the moveable landscape of my childhood. She bought the two red leather chairs at a garage sale near the National Institutes of Health, the Chippendale sofa and Kittenger end tables somewhere in Virginia. And they bought a TV set of their own and sat down every night to watch the latest news about the McCarthy hearings. They were watching on the night of June 9, 1954, when the fight between the Army and McCarthy came to a head, doubly fascinated because it centered on David Schine's efforts to avoid the draft—the same David Schine who came to Vienna with Roy Cohn and destroyed Ted Kaghan's career in an orgy of cartoon patriotism. Both of them felt great satisfaction when the attorney for the Army asked the famous question, "At long last, have you left no sense of decency?"

But my mother also remembered a night when Howard Hunt came over and began defending McCarthy—of course the Communists were trying to subvert the government, he insisted, that's what Communists do—and she got so angry she told him exactly what he could do with Joe McCarthy, an outburst so startling that my father started laughing. "Now you know what she thinks of you," he told Hunt.

Later he told her that he felt the same way, but he wouldn't have said it out loud. It wasn't a good idea to be too outspoken.

By September, my sister was a little more than a year old. She seemed lonely, my mother remembered, so they enrolled her in a nursery school. But half an hour after they dropped her off, the school called and said to come get her—she wouldn't stop screaming.

In December, my mother got pregnant again.

Jocko seemed happy in those years, my mother always said. He played chess and poker with old friends from Vienna, enjoyed his TV and his daughter, did a little fishing in the local trout streams. But he seems to have been very absorbed in his work. When the CIA moved out to McLean and he joined a car pool, she had to go pick him up because the other guys left at five and he always worked till seven or eight. And whenever he traveled to Europe, which was often, she would have to check with Dorothy Hunt to find out how they were doing because Howard wrote every day and my father never did.

Courtesy of Hunt, I know that on one of those trips, they went to Rome to put an end to the CIA's Albanian penetration operations. They had spent years recruiting and training hundreds of Albanians to cross the border and spy or fly over in planes, dropping bags of flour and razor blades and halvah as a hearts-and-minds gesture. But they became certain that someone was leaking their plans after they sent a C-47 across the border to pick up an injured agent and it came under attack from anti-aircraft guns so well placed they fired from two sides. In his memoir, Hunt tells the story of meeting the operation's British liaison at a sidewalk café on the Via Veneto.

> I have forgotten his name, but let us call him Pointdexter. He could have been a prototype for one of Graham Greene's burnt-out cases: With rheumy eyes, trembling hands and capillary-flecked face that had not been shaved in several days, Pointdexter presented neither a professional nor a reassuring appearance.
>
> While waiting our arrival, he had been drinking, and now as we sat down, he ordered a round of double "martoonis" for all of us.
>
> Richardson came quickly to the point, and when Pointdexter had tossed off his drink, he stared unbelievingly at us, eyes welling, tears trickling slowly through the stubble on his face.
>
> "Then it's all over," he said faintly.
>
> "All over," Richardson replied.
>
> "I don't know what they'll say about it in London," Pointdexter said, twirling his empty martini glass.

"London already knows," Richardson told him.

"They *know?*"

"'They know," Richardson repeated, and after some sterile amenities we left, wondering how MI-6 would absorb its alcoholic derelict.

By the spring of 1954, my mother had gained a lot of weight. When her doctor said it was unhealthy to get so fat, she told him that it was probably because she quit smoking. He got annoyed with her. "You can't quit smoking when you're pregnant!" For the last two months she had to stay in bed, so her mother came to help.

I was born in October. Four days later, a hurricane hit Washington and both our oak trees came crashing down, smashing the garage and tennis court. Then the electricity went out and the water stopped and the phone died, and in the midst of the chaos I slept all day and stayed awake all night and threw up whenever I drank milk and cried uncontrollably for hours at a time. The doctor put these symptoms down to colic, which my medical dictionary defines as a mysterious condition that may rise from an extreme sensitivity to the environment.

When my mother was feeling well enough, she went out and bought a beautiful rocking chair with swan arms and rocked me through the night, feeding me formula from a bottle. She remembers my father watching her bathe me in the sink, so proud that he would just stand there and dote through the whole procedure. He even gave me his nickname, switching to the more formal John with everyone outside the CIA inner circle. And she would always remember the night Stephen Spingarn came over—she was expecting the smartest man in the world, the way John always talked about him—and somehow I climbed out of my crib and slithered down the stairs to the dining room! Not even crawling age, and already making a run for it!

So here was John Hammond Richardson after so many years of travel settled in a lovely home in one of those leafy and elegant old Washington neighborhoods. He was forty years old, with a beautiful twenty-nine-year-old wife and a housekeeper, a two-year-old daughter and a baby son. His father's old oil wells were still paying about five hundred dollars a month. As a division chief, he was a rising star at the CIA. Both Dick Helms and Bill Colby would go from division chief to the top.

And then he stepped away, asked to be made station chief again—asked for a demotion. He wanted to go back to the field, to some place that was interesting and politically important. He told Eleanore it looked like Greece or Turkey.

Eleanore said, "For God's sakes, don't send us to Bulgaria."

He got Greece. Once again, he'd be replacing Al Ulmer. But just before we left, the assignment almost fell apart over a minor detail. Allen Foster Dulles asked him to rent a specific house, a large hilltop mansion in Athens that was perfect for diplomatic entertaining. Greece had a royal family and an old-fashioned European court culture complete with courtiers and ladies-in-waiting, so elegance was a professional requirement. But while the house would be paid for by the agency, the parties were another thing, and my father got into one of his stubborn fits and told Dulles he was not going to spend his own money on "representation." He happened to know that Al was practically broke because of the cost of all those parties, and if that's why they were sending him—because it was known that he had his own money—then he simply wasn't going to go.

From then on, we always kept two liquor cabinets, one for CIA booze and one for the family's.

Many of my mother's stories are about transitions, I realize, about the grand public journeys we took from one secret life to another. Nobody went by plane in those days, so when we went to Greece in 1955 we went by ocean liner and it turned out the only berth available was the royal suite, which had a living room and two bedrooms. But poor Little Jocko got such a terrible sinus infection that the doctor came at least four times, and there was a terrible storm and a swamped sailboat and then we switched to an old tub jam-packed with Jewish refugees where you couldn't walk through the aisles for all the people sleeping on the floor. We had to step over bodies as the ship rolled. When the next storm hit, even the dog got seasick—John held him over the rail to vomit and then turned him around.

Then an odd thing happened. We stopped in Naples and John took Eleanore to explore the beautiful hills he remembered from the war, with the magnificent bay spread out below. But when she approached the low stone wall marking the edge of the scenic overlook, John suddenly grabbed her and pulled her back. "Don't go so close." He was breathing fast and sweating, clearly terrified. She was startled. He'd never been afraid of heights before. On their honeymoon in Capri, he drove blithely along those cliffs and cliffs and cliffs.

From that day on, heights frightened him. When he drove over high bridges, he would shake so much, she'd make him stop and get in the passenger seat. He'd reach out as if trying to stop the car with his hand. On the highest bridges, he'd have to lie down on the backseat. Later I learn that psychoanalysts consider phobias a way to control anxiety by focusing it onto a single cause, which makes sense when you add to the loss of his father and his brother all the other separations and anxieties of his unsettled life. And maybe there was a touch of the pull Milan Kundera was getting at when he described acrophobia as "the fear of the desire to fall."

But for my mother, Athens was the adventure of a lifetime. Al and Doris met us at the dock and after the hugs and handshakes, Al told Eleanore he hoped she was rested because there was going to be a big welcome party for them that night. "I can't go to a party," she said. "I have two little kids." But Doris said not to worry, her nanny could take care of them, so they blinked and dressed up and drove to our new house, a magnificent hilltop mansion in the neighborhood where all the royals lived. Built by a Romanian diplomat for his Greek mistress, in a compound big enough for a dozen cars and shaded by thick old trees, it was surrounded by a fifteen-foot white stone wall. The front doors opened on a curving staircase and a grand dining room sixty feet long with a fireplace at one end and a wooden breakfast nook at the other. Near the fireplace, you could push on a panel that opened a hidden door to a secret passage that took you to a huge marble bathroom that was bigger than the average living room and so warm Jennifer and I used it as a playroom. One day my mother asked the diplomat's mistress why the bathroom was so big. She waved an airy hand and said "*pour le monde*"—for the world.

That first week was frantic. My mother supervised the moving and unpacking and interviewed nannies and my father put on his new top hat and went to present his credentials to the king and queen—an event written up in the local newspaper under the headline JOHN RCHARDSON ARRIVES TO TAKE OVER CIA STATION, much to his chagrin. Then my mother decided the grand house was infested with bedbugs and we had to move out and have it fumigated and all the clothes and dishes washed, the first ritual sterilization in a series that would repeat itself in every new country to come.

But soon she began to settle in, and one day she found herself primping in the fabulous bathroom next to Amalia Karamanlis, the beautiful young wife of the man everyone expected to be the next prime minister. They found out they were both the same age and since my father had told her it would be a good thing if she could make important friends, my mother eagerly accepted Amalia's invitation to lunch—especially when she asked her not to park on the street because she didn't want other people to get jealous. So the chauffeur dropped my mother outside the door and they talked about Greek politics and Amalia gave her some poems she had written. Amalia spoke and wrote poetry in German, French and English as well as Greek. But the next time the chauffeur dropped her off, Eleanore admitted she didn't really understand the poetry and instead they gossiped, each promising not to tell anyone what the other said. It would be their secret.

Of course, she told my father everything, and expected Amalia was doing the same with her husband.

Another day she took out the *Vogue Book of Etiquette* to look up the proper procedure for writing a letter, because my father wanted her to meet the princess who lived next door. And just then the phone rang and the butler came rushing in. "It's the princess next door, ma'am, you better come to the phone." Heart beating, she lifted the receiver. "This is Helen," a voice said. "I'd like to ask you and your husband to tea."

So they went to tea and met the king's sister, who told them that she'd broken protocol by calling them because her daughters—Marina Duchess of Kent and Princess Olga of Yugoslavia—were coming to Athens and the Serpieris were giving a party and if she waited to be presented formally, then John and Eleanore wouldn't be able to attend the party. But now they knew each other and didn't have to be presented!

The day of the Serpieri party, Helen called to say her niece and nephew might be at the party too. Not thinking much of it, my parents took us to the beach and ran into traffic on the way home and arrived to find the servants in a panic. They were so late! They had to hurry! They had to be there exactly on time! Didn't they realize that *Their Majesties* were going to be at the party?

Of course! Her niece and nephew! *That's* what Princess Helen meant!

Later that night, she teased my father for his American naivete. "I thought you would understand my cryptic message." But the party was huge and fabulous and my mother would always remember the glass cabinet stacked floor to ceiling with jade *objets d'art*, and before it was over she was able to file away—mentally, of course, because John wouldn't let her take notes—another comic tale for retelling at the millions of parties to come: how the vice-president of Greece came over late in the evening and asked John for a ride home but John couldn't leave before his ambassador so the vice president went to the ambassador and said "Can't you leave?" and he said "I can't leave until the princess goes" and finally the princess left so the ambassador could go so John could take the vice president home!

And that's what it was like every single day. There would be one or two cocktail parties and dinners so formal they often wore tuxedos. And the parties were always interesting. When Sophia Loren made *Boy on a Dolphin*, she came to our house to mingle with a guest list that included Aristotle Onassis and Stavros Niarchos. Then everyone went to Niarchos's yacht, which had a Monet in the stateroom. There were costume parties where everyone came as a famous painting or a song lyric and dinner parties where my mother drove herself crazy trying to seat ambassadors and generals according to rank and seniority. Often she had to consult Lola, the woman who ran the embassy protocol office. One time she forgot to put the French ambassador in the "place of honor" to her right and had to apologize. "It doesn't matter," he said. "Wherever the French ambassador sits is the place of honor."

Another story for the archives.

They sent a Christmas card home that year, the four of us at the Parthenon overlooking the white marble city, my mother glamorous in mink and Dad in his black suit and horn-rims, my sister and I dressed in traditional Greek costumes. I'm sitting in my father's lap, chortling joyfully. "Arrived September 22 and have been very busy ever since. Expect to be here a few years. Sure wish you could come here to visit us! The fishing is wonderful!"

And my father? He spent his first week in the middle of a political crisis, with the prime minister near death and the King still trying to recruit Amalia Karamanlis's husband Constantine to accept the job. There was real concern that the opposition parties might tilt Greece toward the Soviet Union. At one point in the civil war that followed World War II, the Communist Party controlled every part of the country but Athens. It was still active in guerrilla warfare and sabotage, with some five hundred spies running around Greece.

Then the prime minister died and my father put the weight of his station behind Karamanlis, a powerful signal of American support that helped him win a mandate in the Greek parliament. The timing was perfect, giving my father influence right in the heart of the Greek political system. My two best sources on this period are his two deputy chiefs, Lou DeSanti and Gordon Mason. They told me that by 1955 the Athens CIA station was one of the largest in the world, responsible for running operations against the Soviets from Kazakhstan to Hungary. It broadcast free-world news in fourteen languages, maintained its own airport and half a dozen planes and even a few boats. It dropped leaflets all over Eastern Europe and tried to penetrate difficult targets like Bulgaria and Romania and also had vast power over internal Greek politics, almost to the point of running the country. DeSanti described staff meetings so large that ten or twenty deputy chiefs would be sitting around a table. As chief of internal operations, he specialized in fighting the local Communists, building dossiers and organization charts and trying to disrupt their sabotage and espionage operations. Mason was chief of external operations, sending agents into Yugoslavia, Bulgaria, Hungary and Albania and supervising the planes dropping leaflets and spending huge amounts of time interrogating communists who came in across the border. Telling me about this, he insisted that their methods were humane. Once, he said, they turned a female agent over to a psychologist who spent five days with her and almost drove her insane, so they stopped the interrogation and never got the information they wanted. Sometimes he teased my father about not having enough interest in operations because he was so caught up in the world of Greek politics, but the intrigue at the royal court seems to have been effective. Their greatest triumph came when all the opposition parties united against Karamanlis in 1958. Once again, the alliance between the United States and Greece was threatened, which in turn threatened the delicate balance of power throughout Eastern Europe and the Middle East. Working with an unsavory intelligence chief who once tried to stage a coup, DeSanti said, they spent large sums of money in a wide variety of strategies to support and advise the Karamanlis team. On the level of the royal court and the diplomatic world, my father played a particularly conspicuous role. In the end, they helped swing the election, a triumph that earned them a rare special citation from the CIA. "That was a key moment," DeSanti recalled. "We could breathe easily for the next ten years."

Beyond that, they said he had a reputation as a good fellow despite his habit of reserve, that he protected his people, that he focused on the big picture and trusted you to handle the details yourself. Most station chiefs were pushy, but he was the quiet type. He never pounded the table, but worked through persuasion and brain power. He kept up with international issues and writers and often distributed mimeographed excerpts for his staff to read, which they seem to have interpreted as an endearing eccentricity. Sometimes details gave their stories the ring of truth. I talked to one case officer who had polio, a professional disadvantage because it gave him a distinguishing feature. Normally he would have been given a desk job, but my father told him to arrange his meetings with spies so that he was already sitting down. And DeSanti told me that once my father noticed an officer speaking harshly to a Greek, and at the next staff meeting he told the man in front of everyone never to forget that it's their country, that we are guests here. Take this in the best possible way and balance it against his eagerness to manipulate the elections and you have my father, the humble warrior.

Later, in Vietnam, it would have historic consequences.

And there were so many happy days, my mother remembers, drinking and dancing and listening to music and playing golf on that horrible little course way out in Tatoi, the one that had sand greens. Athens was mostly just motorcycles and bicycles then, its beauty unsoiled by modern noises and smells. Gordon Messing had come out to work with John in yet another post and now the three old war buddies all had beautiful wives and children. Sometimes the six of them stayed up all night, enjoying the rare chance to drop their guard in mixed company. Messing took them hiking in the woods above the city. Mason and my father taught the Crown Prince to fly-fish. One time they took the German chargé d'affaires and the Israeli ambassador up to northern Greece near the Albanian border, on a river that flowed right under the Iron Curtain. The Israeli ambassador slipped and his waders filled with water and he floated toward the Albanian border. A sergeant named Pete jumped in to save him.

At the dance that weekend, everyone had already heard the story. The Egyptian ambassador said, "Thank god you didn't invite me or they'd have said I tried to kill the Israeli ambassador." And the Israeli ambassador said, "It could have been the Egyptian ambassador who slipped and they would have blamed me." Then one of the courtiers came up to John and asked him to please ask Queen Fredericka to dance (you couldn't ask the queen to dance unless she gave you permission first) and she immediately started teasing him about the ambassador almost being swept to Albania.

He became such a famous fisherman that one night the queen asked him if he would take *her* fishing. Of course he instantly said he'd be delighted, but the ambassador was the boss and if anyone went fly-fishing, it should be he. So a week later, the queen invited them to lunch along with the crown prince and Ambassador Donovan, and after the coffee came out the queen sat on the sofa next to the ambassador and asked him if he thought it was true that a person shouldn't say no to a queen. Of course the ambassador said it would be quite churlish.

"But Mr. Ambassador," the queen said, "I've asked John Richardson to go fishing with me in northern Greece and he tells me he can't go. Do you think that's right?"

"She's right John," the ambassador said. "You can't say no to a queen!"

So off they went, flying up in a small plane with the head of the queen's staff—her official title was La Grande Maîtresse de la Cour—and another lady-in-waiting and one of the king's equerries and a "gilly" to help the queen net the fish, setting up tents by a stream on the grounds of an old monastery in a town called Yanina. There were personal tents for each of the couples and a huge tent where they did the cooking. Latrines were built. And in the morning they had to wait outside the queen's tent and when she came out, bow and curtsy in their shorts and tennis shoes.

They drove back with the queen. At lunch, someone took a group picture the Grande Maîtresse sent over later, framed in elegant red leather with the royal crown embossed in gold. For years, my parents kept it on the dark cherry table between the two red leather chairs. My mother kept it in her bedroom until she died.

Behind the party anecdotes lay shadows, of course: one day a Polish woman showed up and thanked my father over and over for something he had done, then begged him to help get her daughter out of Poland. But he couldn't. Another time, at a party, a Soviet official came over to talk to him and he turned away, a terrible breach of protocol and waste of a fine opportunity to chat up a Communist. But my father was in one of his stubborn moods. "He didn't like them and that was that," my mother said.

Another time, the ambassador's wife called my mother and asked for Amalia Karamanlis's private telephone number. When she demurred, saying she couldn't betray a confidence, the ambassador's wife put a little steel in her voice: "I'm *asking* you to give it to me." Less than half an hour later, Amalia called. "I understand that you had to give my number to the ambassador's wife. But I want you to know I've changed the number."

Then she added, "Here's the new one."

For my mother, this was a minor triumph. But it stirred my father's old wartime anxiety about his secret status and probably contributed to a small and revealing personal crisis. On birthdays and holidays, when diplomats had to go to the palace through a special door to sign the royal ledger, the Grande Maîtresse would always give my mother a reminder so she and my father could go early. Soon it became known to people like the DCM and the ambassador's wife that the Richardsons always signed the list first, and one day someone at court made a casual remark that "John Richardson is the real American ambassador in Athens." Someone else immediately told the ambassador, who called my father into his office and accused him of exceeding his authority. Or maybe there was more to it than that, because one of his deputies later told me the ambassador asked him for some kind of information and he had to say that he wasn't authorized to share it. What's certain is that my father came home so upset, he sat down in a chair and wept—actual tears running down his face. And when my mother told me the story forty years later, she wept too.

Then two blows hit the CIA. First the Hungarians rose up against the Soviets, attacking tanks with handguns and molotov cocktails, and the CIA cheered them on via Radio Free Europe. Frank Wisner watched from London, convinced that his old dream of sparking revolutions in Eastern Europe was finally coming true. He was now the CIA's Deputy Director for Plans, so covert action was both his passion and his responsibility. At that moment, Israel invaded Egypt on a march toward the Suez, and though the British helped plan the attack and Wisner prided himself on his good relations with British intelligence, nobody tipped him off. So Eisenhower was furious with Dulles and Dulles was furious with Wisner and that was when Soviet tanks rolled into Budapest to crush the Hungarian revolution. Already demoralized, Wisner flew to Vienna and saw the devastated refugees first hand and became obsessed with the irrational idea that Russians were sending Mongol hordes to invade Europe.

Then Wisner came to Athens to stay with us. By the time he arrived, my father recalled, "his motor processes were revved up to an extreme velocity and intensity." For ten days, they holed up at the office, grappling with the Suez situation. At night they sat around the radio listening to the news, disgusted at how we let those poor brave Hungarians down. And each day Wisner got more depressed and tense until finally he came to the house in such a foul mood, taking it out on his assistant in such an ugly way, that my mother got mad and told him he wasn't being very nice.

When he got back to Washington, Wisner checked into a hospital. From this my father drew a characteristic moral:

> The simple lesson of Spingarn and Wisner's example is that one must not consider himself responsible for the whole world nor exaggerate the centrality or importance of the personal role.

Right about this time my own story begins. At first it is just images, moments suspended in the blur of time:

1. A dark room with a glass door opening on a garden.
2. Up in the branches of a fig tree, eating a fig, the red pulp a delicious secret in my hand.
3. Lying on the floor looking at stylized illustrations of men in armor. A woman is reading the book to me.

I am two, three, four. According to family legend, I speak only Greek and my sister has to translate, a phenomenon my parents always seem to dismiss as a bit of childish whimsy.

I am in love with my nanny, Jimmy. I tell her I'm going to marry her when I grow up. I beg her to wait for me. When Jimmy goes out on dates, I'm so jealous I beg her to take me with them. My mother finds this terribly amusing, telling the story over and over. "You'd cry if she didn't let you go along."

This time through the legend, we're sitting in an apartment in New York. Jennifer sits in a big armchair with a pillow in her lap, as hungry for the details as I am. "All I remember is the garden was terraced down and there was a rectangular area at the bottom—that's where I stepped on the bee."

My mother shakes her head, annoyed. "It wasn't rectangular, it was long."

And we ask about our rooms and the secret passage and she takes us back as another mother might tell a fairy story, the time the crown prince stopped his convertible and offered my sister a ride to school and she said no, you drive too fast—how Jimmy cried! And the time I was spanked for stealing figs from the garden, the time we sailed the Greek islands on a beautiful black-hulled yacht called the *Eros*, the time we got chicken pox and they gave us the baby rabbits. She distracted us by painting the walls with characters from *Jack in the Beanstalk*, winding the beanstalk from Jennifer's room to mine.

Jennifer wants to hear the deer story, one of her favorites. Mom was walking with a friend through the woods on Niarchos's private island when a crazed buck attacked them, rolling her friend down the hill with his horns, and somehow Mom managed to jump on its back and strangle it to death. This story always seemed wildly improbable to me but Jennifer loves it and so did Dad, smiling proudly whenever it came up. "Later Niarchos offered to give her the horns," he would say.

Mother would shudder. "No, thank you!"

And what about the dresses? John's secretary was dating a dress designer and he designed so many beautiful gowns for both of them.

"That was George Stavropoulos. He was quite a famous Greek designer."

"Didn't Dad give you a jewelry fund too?"

"He gave me a jewelry fund," she says, covering her face with a hand. "You're making me cry."

When it was time for us to move to a new post, the king and queen invited us for a goodbye outing at Piraeus harbor. This is the last of my dreamy early memories, the white sails on the water and the important adults sitting together as I slipped down to the shoreline and scooped up a handful of those little river shells shaped like miniature horns of plenty. I remember everything about it, the damp little shells and the muddy marine smell and the delicious fear of getting caught. Then I said goodbye to Jimmy and we sailed out of the same harbor, a dislocation that put me through a dramatic change that still seemed to annoy my mother forty years later. "You started trying to speak Greek to all the waiters on the ship and that didn't work, so suddenly you were speaking English with a Greek accent—because you wanted something."

We were heading south, aiming for the Suez Canal and the Indian Ocean. Impressed with his handling of the royals, the Agency had offered my father London, another prestige post full of pomp and circumstance. My mother begged him to take it. But he said it was "just liaison."

They offered him Switzerland.

"Too boring," he said.

The Communists were on the move in the Far East, so that's where he wanted to be—where the action was.

The trip to Manila was another one of my mother's set pieces, how our cruise through the Suez was cancelled because of a strike and the luggage wouldn't fit on a plane so we had to wait a couple of weeks for another ship, killing time with a side journey to ride camels in the Bekaa Valley—and I was so outraged, she always told me, when the camels spit at me. "*'That's not polite,'* you said. You were very put out." We spent a week in Tehran with the Hansens, our friends from Vienna. Ken was an advisor to the Shah. We saw the Peacock Throne and also the terrible poverty. "Remember how that woman kept her face covered with her chador but lifted her skirts to pee in the ditch?" Then we went to India where all the men had turbans and beards and there were vultures sitting on all the trees. We rode elephants. We went to the Taj Mahal. And there was a hotel in New Delhi with a big marble balcony hanging over a pool that was filled with ducks. "You jumped right off it into the water. You swam with the ducks."

In Burma, my father put a deep wound in his foot hiking to the golden pagodas.

In Bangkok, they did temple rubbings and watched rats jumping through the weeds.

In Hong Kong, they ate at a floating restaurant near Repulse Bay and my father got a visa stamp in his passport that identified him as "a foreign service reserve officer."

A few days later we arrived in Manila to heat so awful we spent the first few days "panting like mutts." It was dusty and ugly and the streets were a stinky chaos of buses and "jeepney" taxis. When we got to our new house, my mother's heart sank. It was in a "colorful" beach community called Paranyaki but only the bedrooms were air conditioned and it was overrun with so many cockroaches and mice and ants that even after extensive fumigation, she had to get glass jars with screw tops for all the rice and pasta. Before long we moved to Forbes Park, Manila's most elite neighborhood, not far from the Polo Club. With palm trees in the front yard and a large outdoor wooden dance floor in the back, this house was elegant enough to satisfy both my mother and the embassy. My sister and I got a new nanny, a Filipino woman named Mercy who would keep us in line by slapping her shoe against her thigh. We started spending our days at the Polo Club, swimming and riding horses. And again, we were surrounded by history. Sitting at the embassy you could look across the bay to Corregidor Island, where the Americans made their last stand and MacArthur said "I shall return." Just beyond that was Bataan, site of the infamous Death March.

Settling in, my parents went to their first formal party for the president. They were amused by a sign at the entrance: LEAVE ALL FIREARMS AT THE DOOR.

In fact, the Philippines was a troubled and important place, and my father was there for a reason. Strategically vital because it straddled the shipping lanes of the Far East and hosted vast American military installations like Subic Bay Naval Base and Clark Air Base, politically significant as the "showcase of democracy" in Asia, it was also plagued by an exceptionally greedy ruling class that created social unrest through its relentlessly unfair treatment of the poor. During World War II, as rich Filipinos collaborated with the Japanese, the Hukbalahap "people's army" fought alongside United States soldiers. Afterward they went to the ballot boxes and elected their leader—a former tailor named Luis Taruc—to the Philippine

Congress. In retrospect, they seem to have been less interested in revolution than fair treatment, but the United States was eager to sign a trade agreement and afraid of communism and chose to side with the rich, standing aside as the Philippine government forced Taruc and the Huks back underground. The result was open rebellion. Within a few years, the Huks controlled much of the countryside and even parts of Manila. Then a CIA agent named Edward Lansdale groomed an effective and decent leader named Ramon Magsaysay and helped him defeat the Huks, both by military means and by reaching out to them with political reform and an innovative surrender program that gave each guerilla a piece of land and a water buffalo, the beginning of the "hearts and minds" concept of counterinsurgency. Lansdale also wrote Magsaysay's speeches and financed his presidential campaign, which was either (depending on how you look at it) illegal manipulation of a foreign government or one of the CIA's great triumphs.

Then Magsaysay died in a plane crash. The new president was corrupt and unpopular and by the summer of 1959, as a CIA officer named Joseph B. Smith said in his memoir, "We were witnessing the total collapse of the showplace of democracy that had been the pride not only of the CIA but the Eisenhower Administration." Even so, people at the embassy were surprised when we arrived. According to James Lilley, a prominent future ambassador who was then working for the Manila CIA station as a lowly China specialist, people at the embassy puzzled over it. "Your father had the reputation of being a very high-powered guy who did a great job in Greece manipulating the royal family," he told me.

So what was he doing in the Far East?

The reason soon became clear. As we settled in, Lilley said, my father became known as a driven man who believed the CIA could change the world. He had three main interests. The first was penetrating the Communist Party. At the time many of Manila's leading intellectuals had Communist ties and well-known local Communists seemed to be able to live underground indefinitely, but the station had no agents inside the party.

Second, he wanted to forestall a return to warfare by cleaning up the remnants of the Huks. So he brought Lou DeSanti over from Greece as a counterinsurgency specialist. By way of a welcome gesture, he spent a Sunday afternoon driving DeSanti out to Bataan in his black diplomatic car. Later DeSanti joked about tracing the Bataan Death March by chauffeur, but there was no mistaking my father's sense of mission. Quickly he went to work with Filipino intelligence and military, helping them with training and money and "all kinds of things," DeSanti said. Soon they started capturing high-level Huk commanders like Casto Alejandrino, Taruc's right-hand man. When they caught Commander Juan Sumulong, leader of a remnant gang of Huks who ran prostitution and gambling near Clark Air Base, DeSanti remembered my father asking the Filipinos not to torture him, although he insisted it wasn't motivated by mere decency. "What we learned in the Philippines was this: you cannot confront a guerrilla movement successfully unless you have the people with you. It's the most important lesson in counterinsurgency."

But my father's third interest was more unusual. Unlike his predecessor, he had little desire to send agents across the Chinese border, figuring the most they'd do was send back a few reports on plane traffic before getting caught. Instead he pressed his team to develop better contacts with the leading social powers, like the Jesuits who controlled the church and the wealthy Chinese who ran the economy. Even after four years of watching my father work the Greek royals, DeSanti was surprised at how intensely he pursued local bankers and economists and college professors, inviting them to dinner at our house and "activating" them in all kinds of ways that went beyond ordinary espionage. "I've never heard of any station chief doing that before. It's not one of the missions—to fight against Soviet communism, local communism, yes. But to activate the elite, that's not stated. It was a marvelous thing to watch."

Most of all, my father wanted to find a new Magsaysay. Immediately after we arrived, Smith got an invite to a secret meeting with the vice president. In his book, you can almost taste his pleasure at how simple it was to get the go-ahead: "Jocko said, 'Go meet him.'"

This is what I remember:

4. I am in a new school with outdoor walkways and a bamboo forest and all the other kids are brown. I like the green secrecy of the bamboo. I love my new Pental pen, with its soft tip and exciting scientific smell.
5. The desk is wooden, shared by two boys. I don't know where to put my elbow. The nun stands over me, annoyed.
6. At night, I dream of walking into class and looking down

to discover that I'd forgotten to wear pants. I dream of getting lost.

It's odd that my free-thinking father put me in a religious school. My mother liked to tell the story of their interview. "You understand that if you put your son in a Catholic school," one priest said, "we'll make a Catholic out of him."

My father chuckled. "Go ahead and try—and we'll try to make sure you don't succeed."

Later I learned that Ateneo de Manila was the Harvard of the Philippines, a significant power center with ties to the government and influence over public policy. My mother said the priests came to our parties and knew my father so well, a couple of them even came to visit us in Washington.

I Was a First-Grader for the CIA.

But not much of one. I came home crying the first day because the kids called me "a little white monkey."

Another day I came home crying because the nun kept hitting me with a ruler for writing with my left hand. Whack!

I told her I was left-handed.

Whack!

I got Cs for conduct.

Finally my mother went to the principal and demanded they let me write with my left hand. Sister Loreto never forgave me.

Again, protocol was my mother's obsession and nightmare. On our arrival, the political counselor's wife came over with the diplomatic list and drew a line through it to separate the people my mother had to call on from the people she could receive. Later Ambassador Charles Bohlen's wife held a meeting to give the wives protocol lessons, and one woman grew fretful because her house was full of sofas and Mrs. Bohlen said it wasn't lady-like to sit on a sofa. It got worse when Bohlen was replaced by John Hickerson, whose wife wore a Phi Beta Kappa key around her neck and terrorized the embassy wives. Thinking it would encourage relationships with the local community, she made a rule that Americans couldn't fraternize with Americans, enforcing it so strictly that even when an old friend came to visit from the United States, you couldn't invite other old friends over to see him. Once my mother gave a particularly glittering bash that included Americans along with the Filipinos and Chinese, and a gossip columnist ran some incriminating pictures in the local paper. The next day Mrs. Hickerson called the DCM's wife and said she wanted to see my mother first thing in the morning.

My father got in one of his stubborn moods. "Absolutely not," he told her. "You will not be called on the carpet."

When word of this got around the embassy, the political officer called my mother. "Are you really not going to go?"

No, she said.

"I'm really proud of John," he said.

Mrs. Hickerson never spoke to my mother again.

Then I come awake again.

7. On a shady street with a canopy of tall trees, I hop up on the stone wall. I'm walking at my father's level, pursing my lips and blowing pffts of air. He gives me pointers and then the pffts turn into chirps and then I'm whistling— whistling!

8. I call his office, excited to be connected by the magic of the phone to his important world. I want to ask him if Wednesday is really spelled W-e-d-n-e-s-d-a-y when any reasonable person can clearly see it should be W-e-n-s-d-a-y. His secretary tells me he is too busy but that it is W-e-d-n-e-s-d-a-y.

9. There are horses and a girl named Aurora and our parents are somewhere over by the clubhouse, doing something.

10. The servants are cleaning the dance floor. Our nanny tells us to watch out for banana spiders, they hide in the clumps of fruit and bite you when you reach up. Soon we will perform for the guests, showing off our bows and curtsies.

In the stories my mother tells, my father's preoccupation with his work is a running gag, the flip side of my perverse insistence on speaking Greek. Like the time he took me back to Manila by himself because my sister had to stay in Washington for a minor operation. At the airport, saying good-bye, she told me to make sure my father didn't forget his bag. But I must have looked away because he ended up getting on the plane by himself, without the bag or me either. "Then I saw you running through the gate with both bags, yelling, 'Dad, you forgot me!'"

But it didn't bother me, my mother insists. "You were happy as a clam. Everybody on the plane made a fuss over the poor forgotten boy. You were the darling of the crew."

Then we stopped in Tokyo and went for a drink with the CIA station chief and they went back to the office, too absorbed in their conversation to notice that little Jocko wasn't behind them.

And what absorbed my father so completely? He played tennis at the Army-Navy Club during his lunch hour, no matter how terrible the heat. There were many luncheons and dinners with waiters in white jackets and garden seating in rattan chairs. There were large formal parties when my sister and I helped polish the silver while the servants unpacked my mother's party treasures—the Steuben crystal, the Limoges china, the Meissen vases—then joined our parents on the reception line and bowed and curtsied and served canapés. My father loved his wooden dance floor, developing a reputation as a wonderful dancer who was particularly skilled at waltzes and always charming with the ladies. And he kept up his read-ing, ordering books from Brentano's in New York; among his papers, I found a receipt for *The Road to Wigan Pier* by George Orwell and *The Search for Good Sense: Four Eighteenth Century Characters*. On Sundays we went to the Polo Club with DeSanti and his three kids and also our new civilian friends, an embassy doctor named George Mistowt and a businessman named Herbert King-Heddinger. Sometimes we went up to the mountain resort of Baguio, six thousand feet up in the mountains, where there was a lousy little golf course with sand greens.

The rest of the time, he worked. In his memoir, Smith gives glimpses of the operation side of the station, of tapping the phone of a black market operator and meetings with colorful Filipino politicians like the mayor of Manila, who kept a .45 in his belt and had "Chinese tea" with a new woman every day between four and four-thirty. But my father kept his secrets to himself, as usual. He never even mentioned Smith's book, though Smith calls him "the wisest, most supportive man I ever worked for." Most of what I know about his professional life, I learned from his coworkers. Once a year, he hosted a meeting with the station chiefs of the Far East Division. Occasionally, he'd send a man into China. He developed a relationship with the country's leading cigarette smuggler. He used George Mistowt's house to have a secret meeting with a young Navy lieutenant he was trying to recruit. He bought a small printing press and published bootleg copies of anti-Communist literature like Salvador de Madariaga's *Anatomy of the Cold War*. He was part of the festivities when Eisenhower came out in 1960 to meet President Carlos García. At one point, he was excited about a left-wing Filipino whom Lilley and Lou DeSanti were cultivating, hoping he would finally give them a line inside the Communist Party. It was a long process of what they called "officer ascendancy," inviting the man to the nice house and the impressive dinner parties and wooing him with subtle conversations that showed the depth of their understanding and conveyed the important benefits that cooperating with the United States could bring.

But the big task was still the same: to find a new Magsaysay. The rising star of the Liberal Party was Ferdinand Marcos, a senator whose wife Imelda was widely known as a singer and famous beauty. She tried to make friends with my mother, sending her a beautiful blue dress with butterfly sleeves. Another time they invited us to Baguio, where I spent the day playing with their son Bong-Bong. My mother liked Imelda and always said she seemed very nice, but my father discouraged their friendship. He told her he thought Marcos was a murderer. He preferred the vice president, a relatively liberal Catholic named Diosdado Macapagal, author of many books with titles like *The Common Man* and *Building for the Greatest Number*. So she became close to Eva Macapagal instead.

In his memoir Smith says he argued against Macapagal, trying to convince my father that under the reformist talk he was just another deal-making politician. A political liberal who believed passionately in the "hearts and minds" nation-building approach that stopped the Huk revolution, Smith preferred a candidate named Emmanuel Pelaez, a liberal who really seemed to believe in a Philippines "free from economic and political bondage to a small, self-perpetuating elite." But Macapagal brought Marcos into his coalition. You couldn't trust a man willing to play politics. He quotes my father's response at length:

"Let me tell you something," Richardson said. "I went to a college in California with a man just like Macapagal. The college was Whittier, and my classmate was a man named Richard Nixon. Look, Nixon was running for President of the United States even then.

"Nixon wanted to be president of the student council at Whittier. All the successful candidates for that office had always come from a jock-strap fraternity. Nixon reasoned he had to make that fraternity as step number one. So he went out for football. Hell, he couldn't begin to play football, but even though he fell all over himself, he managed to make the scrub team. That got him into the fraternity. And he got elected president of the student council. If you're in politics, Joe, you have to be a politician all the time."

Smith wasn't convinced. And he despised Nixon. So when the CIA gave them two hundred and fifty thousand to spend on the senatorial elections, he used all of his persuasive talent to swing the money to the Pelaez ticket. Finally my father agreed to give Pelaez two hundred thousand, a remarkable bit of deference to a subordinate's judgment. But he insisted on giving the last fifty thousand to Macapagal. That way, "if Macapagal and his Liberals won, we would still have the prospect of a change of administration led by a friend."

Macapagal and the Liberals trounced Pelaez, turning my father's fifty thousand hedge into money very well spent. Soon he and his men plunged into a series of meetings with all the new senators and behind-the-scenes powers, plotting to consolidate their influence and manipulate the presidential election of 1961—to get rid of the corrupt and unpopular President García and replace him with my father's choice, Diosdado Macapagal. As to Macapagal's devil's bargain with Ferdinand Marcos—well, they'd just have to live with that.

I took papayas around the neighborhood in my little red wagon to sell them.

At my sister's birthday party they found cobras in the wall. They caught them with a loop and pulled them out, all the kids watching with big eyes.

The boys at school convinced me that I was going to go to hell because

I'd never been baptized. I badgered my father about it, getting whiny and insistent in a deliberate attempt to irritate him.

> 11. He comes into the bedroom with a crystal cocktail tumbler in his hand. He's angry or annoyed and there's a compact forceful power coming out of him like heat off a stove. He splashes water on me and says, "There, you're baptized."

Then there was *Mighty Mouse*. Enrique had a TV and we'd watch it every week and go out and play on his swing set. One day his father came back from a trip to Spain with some black candy he called "regalese," a potent licorice that tasted dense and elusive and a little bit wrong, like sugar gone bad.

One day Enrique leaned over a barbecue pit at the beach just as his father squirted gasoline. They chased him and rolled him in the sand but the sand got into his skin. At the hospital I remember high ceilings and green halls, the smell of alcohol and gauze. The chewing gum I gave him was the first thing he ate. I read to him from my book of Greek myths. Later my mother said his face was burned too badly for bandages, but all I remember is the book of myths and the smell of alcohol and gauze.

Another day I was alone in the house and decided to make a pair of wings, certain that I was so remarkable and close to magic that I could jump off the kitchen roof and fly. So I found a big cardboard box and a pair of scissors.

> 12. I hide in a corner of the yard and poke a hole into the cardboard, sawing up and down until . . .

The rest I don't remember, but the scar is still there, the size and shape of an almond. Mother said the scissors slipped and I hit an artery, the blood gushing like an oil strike. "Mercy put a tourniquet on you right away. She saved you."

In the 1960 elections, my father supported John F. Kennedy.

In March 1961, the American invasion of Cuba crashed into defeat at the Bay of Pigs—the CIA's most public failure, an inept and foolish disaster that demoralized my father and all his friends. Immediately afterward,

Kennedy fired both the head of the CIA and the Deputy Director for Plans.

Eight months later, the Filipinos elected a new president: Diosdado Macapagal.

On the Magsaysay campaign, Lansdale had used dirty tricks, like drugging the drink of the opposition candidate to make him seem drunk when he gave a speech. It's very hard for me to picture my courtly father authorizing anything like that, but I really don't know the details—just my mother's story about the immense victory parade when we were up on the reviewing stand, covered by a big awning as the troops marched past for hours and a Jeep went by with three MPs and a dog and all of them saluted, even the dog. Of course it was very surprising for us to be there at all, a CIA chief standing right next to the president. "But John was said to have done a lot to help Macapagal," she said.

By whom?

"Don't ask me. But it was said."

Macapagal turned out to be a fairly good leader. But in the next election, perhaps because his deal-making tendencies backfired on him, perhaps because the new CIA chief chose not to manipulate the elections, he lost to Marcos—who for the next twenty years methodically squashed all hope of a Philippines free from bondage to a small, self-perpetuating elite.

In the early months of 1962, my father flew to Baguio for a conference with Averell Harriman, Kennedy's new Assistant Secretary of State for Far Eastern Affairs. Vietnam was sliding steadily into a crisis and the CIA needed a new man there, someone steady and experienced in both insurgency and war. But Harriman and my father did not hit it off. At seventy, Harriman had run a bank and a shipping company, negotiated with Trotsky and Stalin, and served in Truman's cabinet. He was famous for his high regard for his own opinions, which included skepticism about the Cold War and cold warriors and a growing belief—correct, as it turns out—that Communism wasn't monolithic and Vietnam was a bad place to take a stand. And when he reached a point when he didn't want to listen anymore, he simply reached up and turned off his hearing aid.

Telling me about the meeting later, my mother remembered how annoyed my father was when he came back to Manila, fuming at how arrogant Harriman and his young aides were.

But the CIA offered him the job of chief of the Saigon station anyway.

He was "tickled pink," my mother remembered.

Before he left, he wrote a memo to CIA headquarters saying that he didn't see how we could win the war with that long porous border along Laos and Cambodia—a border just seventy miles from Saigon, far too long to seal or defend. He got a cable back from headquarters that said: "Proceed to Vietnam, and leave the global thinking to us."

We left in June.

21

•

SAIGON

Saigon was warm and humid, a romantic place with French boule-
vards and palm trees and beautiful women in sleek dresses. "A provincial
French town with coconut palms," one writer called it. But the war was
heating up. As the American helicopters started to arrive, the rents on
some of the nicer villas shot up to a thousand dollars a month. At the La
Cigale nightclub, reporters and soldiers and spies crowded around a beautiful Vietnamese singer who crooned doleful French songs. The Rue
Catinat teemed with pickpockets and prostitutes, moneychangers and food
carts. Sometimes you heard gunfire.

Once again we started out in an awful little house with millions of
giant cockroaches, but soon the wife of a CIA officer named Lou Conein
helped find us a fine large house at 38 Phung Khak Khoan, also known as
Ex Rue Miche by those who were slow to give up French colonial names.
I remember going to see it for the first time; a guard shack at the end of
the driveway, big empty rooms with tile floors and thick plaster walls. I
ran upstairs to a brown wood door and then stepped into my fresh

new universe, a room with a big corner window and a desk built right into the wall. The bed was built into the wall too, with a storage compartment above that had excellent possibilities as a hiding place.

Meanwhile, my father was settling into the small office on the second floor of the American mission with a brass plaque on the door saying "Special Assistant to the Ambassador." He spent much of his time flying around South Vietnam with Bill Colby, the CIA chief he was replacing, visiting training camps and mountain tribes and the controversial fortified villages called "strategic hamlets." Sometimes he traveled with Rufus Phillips, a former CIA man who came out to organize the strategic hamlet program for the Agency for International Development, and later Phillips told me about the feeling of hope they had in those early days, when he and his team were distributing hundreds of thousands of school textbooks and training teachers and organizing surrender programs and even replacing the little pot-bellied Vietnamese pigs with fat American swine. They were convinced that the strategic hamlets were the only way to avoid a conventional war and buy South Vietnam enough time to survive.

The core problem of South Vietnam in those days was the president, Ngo Dinh Diem. So pious that he kept an oath of chastity his whole life and prayed several hours every day—his enemies would joke that he became a politician because he found the church too worldly—Diem rose to political power in the early 1930s and then quit to protest the French occupation. Later Ho Chi Minh asked him to join the rebels but he refused that too, fleeing to a monastery in the United States. When he returned in 1954, he took over a country swamped with nearly a million refugees and reeling from power struggles between religious sects and criminal gangs and rogue army generals. Within six months a gang of river pirates called the Binh Xuyen was pounding his army on the streets of Saigon and an American general named J. Lawton Collins was already wiring his obituary to Washington: "I see no repeat no alternative to the early replacement of Diem." But Diem surprised them all, crushing the gangsters and resettling the refugees and consolidating his power. Over the next few years he increased agricultural production by 25 percent, built a railroad the length of Vietnam, and even tried a little land reform. But his virtues cast long shadows. Having overcome so many opponents, he got in the habit of putting his critics into prison. He refused to allow elections and took a high-handed approach to peasants: he was the father, they were the children. He hunted and killed the Vietcong rebels, still widely viewed as the great nationalist heroes who drove out the French colonialists. He turned inward and gave increasing power to insiders like his younger brother and chief political advisor, Ngo Dinh Nhu, often described as a vain and arrogant schemer fascinated by totalitarianism and power, close to insanity and possibly a drug addict. Another favored insider was Nhu's controversial wife, the infamous Madame Nhu, a fiery beauty who had survived a Vietcong prison and often pushed Diem to take tough stands. She was the one who goaded him into defying the river pirates, for example. Later she took command of her own private militia of uniformed Amazons and pushed an unpopular series of decency laws that banned adultery, contraceptives, prostitution and dancing. By the time the Vietcong launched the wave of terrorism that announced their new guerrilla war, murdering hundreds of schoolteachers and decapitating more than a hundred village leaders, the intellectuals and urban elites had turned against him and his army was too weak and demoralized to fight. By the summer of 1960, CIA analysts were already predicting the end.

That November, hundreds of rebellious South Vietnamese paratroopers surrounded Diem's palace and demanded a new government. The United States held back, waiting to see what would happen.

But Diem survived again, and soon Nhu went back to arresting and purging and filling the prisons. Madame Nhu's pet newspaper began making references to "foreign people who claim to be our friends."

A few months later, Kennedy became president. After studying the situation, he decided to try to repair the relationship with Diem and sent out a new ambassador named Fritz Nolting, a soft-spoken Southerner with a doctorate in philosophy. He put the military under a new commander, General Paul Harkins. He gave both men orders to encourage the Vietnamese and give them moral support. After an influential State Department staffer named Roger Hilsman argued that the only way to defeat the guerrillas was by adopting guerrilla methods, Kennedy also boosted the CIA, quietly giving the Saigon station another million and a half dollars and forty new staffers—and a new station chief.

Many years later, I found some notes my father made for a CIA briefing. He arrived in Vietnam, he said, with "3 injunctions & a personal feeling."

1) Turn over of major para-military programs.
2) Intensive coordination with MACV.
3) Roger Hilsman and guerrilla effort.
4) My views on strategic importance of Laos.

The major paramilitary programs included a detachment of four hundred U.S. Green Berets under his command. Currently they were assigned to help train the Catholic militias (#3) and the Montagnards, the primitive mountain tribes that lived in the jungles along the border with Laos (#4). Putting the Green Berets under CIA command gave Kennedy a way to support Vietnam without getting too deeply involved. "It was one of the most successful programs for using civilian forces ever devised," according to a subsequent Army study.

Now he was supposed to turn that over to the military. "Operation Switchback," they called it.

But the most controversial entry on my father's list was item #2, his desire for "intensive coordination" with the military command. According to the CIA historian who wrote the Agency's classified internal history of this period, my father began his tour as Saigon station chief with a formal announcement that he was going to bend CIA resources to supporting the military effort. A dramatic change from past CIA practice, this was probably a response to both the debacle at the Bay of Pigs and the escalation to full warfare. Operation Switchback was certainly a factor. From one point of view, it was a reasonable reaction to changed conditions, a gesture of modesty and teamwork.

From another, it was a big mistake.

His most important relationship began with a classic bit of CIA back-alley intrigue. That first week, with Bill Colby at his side, he slipped through the back gate of the presidential palace and up a flight of stairs to a long soundproof office decorated with books and stuffed animal heads. There he met the CIA's direct line to Diem, Ngo Dinh Nhu. For the past four years, Colby had been making this trip every week. Lately the relationship had cooled and now it was my father's job to warm it up again.

Nhu was thin and intense, a chain-smoking intellectual with a degree in literature from the Sorbonne. If that first meeting was anything like the ones Colby described, he would have gone into long tangents about his past struggles and political theories and methods of tiger hunting while the cigarette smoke curled around him and servants slipped in to empty ashtrays and refill teacups. My father's French was crucial, since Nhu didn't speak English. They must have shared a few memories about their student days in France.

As they got to know each other, my father found much to admire. Nhu seemed cultured and very intelligent, far more absorbed in the details of government than any mere schemer would be. Yes, he spoke in three-hour monologues, but my father didn't let that bother him as much as it seemed to bother other Americans. Yes, Nhu seemed to believe that the only way to fight the Communists was to adopt some of their methods, creating a web of secret police and a huge private paramilitary. But Vietnam was a country at war, and Nhu also had to contend with the ambitious generals and courtiers who were constantly plotting against his brother. He did seem to be geniunely committed to the strategic hamlet program.

Nhu had a grand scheme. He wanted to use a mystical religion called Personalism to meld the villages into fighting units. Personalism would be a third way between communism and capitalism, teaching peasants to find individual fulfillment in submitting to group discipline. Only with time and development would something closer to Western freedom be practical. To most Americans on the scene, this seemed like an attempt to create a Fascist cult. But my father was intrigued and even enthusiastic, I am told. Maybe he saw it as a kind of Asian stoicism.

At the very least, it showed that Nhu was trying.

This is what I remember.

13. This school is really just a rambling old house with chickens pecking in the yard. It smells like soup. Trees with big leaves cast a greenish shade, and everyone speaks French.

14. I'm in the locker room of a big old wet building and the grown men are walking around, ignoring me. Outside there's a big pool with wide gutters splashed by little waves.

15. There's a monkey next door that perches on the wall. I climb up and sit with the monkey and everybody makes a fuss. Then it bites me.

16. I like hiding in trees, in the center of the green sphere. Nobody ever looks up. You can sit five feet above their heads and they never know you're there.

Beyond that, most of those early days are a blur. I know that my mother didn't have to make and receive calls because it was a war zone, that she wanted work done on the new house so we had to wait in the little house until it was ready, that she couldn't find any acceptable furniture so she found someone to build some and flew to Hong Kong to buy fabric. While she was gone our Filipino nanny—brought along for continuity, like the books in the library—kept us busy with swimming and tennis lessons at the Cercle Sportif, where Colonel George S. Patton Jr. taught my sister to dive. My sister made friends with a girl named Doris Meryn and they spent all their time playing Barbies and going to Elvis Presley movies (but listening to Pat Boone).

One night, my parents went to see how the work was progressing on the house at Ex Rue Mich and discovered that the night watchman had abandoned his post. On investigation, my father learned that the watchman left every night as soon as the sun set because (he explained) his dog didn't like the ghosts that came out at night. Then my mother found out that our new servants didn't like the ghosts either. Apparently the house had been used by the French intelligence officers of the Deuxième Bureau before passing to Japanese military intelligence and then to the Vietcong rebels and back to the French, and the servants (and his dog) were convinced that the souls of all their tortured prisoners lived in the series of little rooms in the backyard. So we hired a team of Buddhist bonzes—the local term for monks—to chant and burn incense and hang mirrors over all the doors. Supposedly the ghosts would see themselves in the mirrors and run away.

I nailed a crucifix to my door. "Those mirrors might be good enough for Buddhists, but I'm not taking any chances!"

I don't remember that. But I do remember going with my mother to look at rugs because I was sitting near a shop window studying a tiger skin when suddenly I had that feeling of waking up again, coming alive to the knowledge that I was stuck, bored and helpless, towed behind an adult whose purposes had nothing to do with me. I found myself tugging at the tiger's paw. One of its claws was a little loose. It was so glamorous, so wickedly curved, I started trying to work it free. Fear and excitement rose beside a sneaky calm. I glanced at my mother and saw that she was still talking to the clerk, ignoring me completely, and I gave the claw a last tug and slipped it into my pocket—the beginning of my criminal career, conciousness born in a moment of larceny.

As we settled into our new house, my mother walked the dogs at night just like life was normal and guns weren't firing at night in the delta just outside town. She always did it hours before the midnight curfew and Dad didn't object, but then the security people at the embassy found out and that was the end of that. She kept busy with various projects, working with Lindsey Nolting to raise money for an orphan hospital and taking art lessons from a talented Navy wife named Kay Drachnik, painting landscapes and still-lifes in Kay's garden. She spent a few mornings a week trying to teach a neighbor's child to talk. He had Down's syndrome, but eventually she got him to remember a few words.

My father kept busy. Sometimes he flew up to Da Nang to visit the training camp where the Green Berets taught the mountain scouts and the trailwatchers, or up to the highlands to visit the Montagnard villages, and once he brought back a couple of short wooden spears that looked like something from one of my Greek books of myths. That's when someone gave him a gun as a gift, a sporty little snub-nosed .38 with JHR carved into an ivory handle. He usually came home late, sometimes at seven but more often after ten. Sometimes other CIA men or Vietnamese officials would come over and we all had to go upstairs so they could talk privately.

On weekends he played tennis and golf.

I remember one day vividly. It starts in the bathroom, Mercy standing behind me and pushing my bangs up into an Elvis wave with her palm. She rakes the comb hard across my skull, letting me know she's annoyed with me. In the mirror my nose seems too small, my lips too big and girlish. Then I'm in the garden and the adults are talking about a new dance craze called the twist, which is so popular—or so I'll learn years later—that people are defying Madame Nhu's ban on dancing by sneaking off to private nightclubs they jokingly call "twisteasies." Suddenly I decide that since I'm a kid and a member of this crazy new kid generation, I should demonstrate the twist, even though I've never seen anyone dance it and have no idea how it's done.

17. Now everyone is watching me, even the monkey is watching, and I'm flailing around and praying that I look amazing and suddenly I catch a glimpse of my father's face—and he's frowning, embarrassed and disgusted. I have no idea how to do the twist! I'm a fraud!

Another day, I'm in my mother's room and she is getting ready for a massage. The dressing table is a little theater, its stage set with her silver hairbrush and mirror and red leather jewelry box. It just rained and the air is cool and fragrant. Then something makes my mother glance at me with a strange expression, as if she's just noticed something.

18. "You're going to be a real lady-killer when you grow up," she says. It sounds bad but she doesn't seem angry, just distant and objective. When the masseuse comes I wander outside and slip past the guards into the street, cut down one side street and then another until I arrive at a street market: dried garlic hanging in bunches, open baskets of grain, big glass jars of dried fruit. The air has a rich spicy smell.

But that's all. Sometimes I wonder if the memories slipped away because there was nothing to anchor them, no familiar places or old family rituals. My sister feels the same way. Wistfully, she reminds me of the times our parents took us to the Brink BOQ so we could eat on the rooftop overlooking the trees and the city and the river, the exotic sight of steaks grilling over open coals. "They would smear butter on both sides of the steak and they'd slap it on the grill and it was the best thing I ever tasted. That's one of my strongest memories."

"I remember fireworks," I say.

"Not fireworks," my mother snaps. "Bullets. Rockets. There was a war on, you know."

22

•

A PLANE CRASH, A BOMB

By September 1962, the Vietcong were attacking a hundred different places every week and killing three hundred Vietnamese soldiers a month, and news reports out of Saigon were increasingly gloomy. How was a priggish and unpopular dictator like Diem going to beat the heroic guerillas who beat the French? But the U.S. Army and embassy kept saying that everything was going just fine. After so many American near-betrayals, the job of giving moral support to the embattled regime seemed to require a constant stream of happy talk.

That was when a young reporter named David Halberstam arrived in Saigon on assignment for the *New York Times*, going directly from the plane to a goodbye party for a reporter who was being expelled for saying the war was a losing proposition.

The next day, he had lunch with my father and came away with a good lead. So he asked another reporter if that sort of thing happened often.

"No," the reporter said. "They're making a play for you."

A few weeks later, they had lunch again. Halberstam recalled the meeting in a book called *The Making of a Quagmire*: "Our talk was pleasant that day, but peculiarly involved a long, abstract discussion on the nature of counterinsurgency. I was never quite sure that I understood Richardson correctly, for our terms and points of reference were so different."

Most likely, my father was talking about his experiences with the Hukbalahap in the Philippines and the hearts-and-minds efforts they were making up in the mountains, where Green Berets were digging wells and repairing schools and stocking ponds and even distributing school books and blackboards. It's the kind of thing you'd tell a journalist, and it had the additional virtue of being true. Not only was the hope geniune, but the counterinsurgency programs were probably America's best chance to stay out of the shooting war.

But Halberstam wasn't buying. And things got worse when he started asking questions about Ngo Dinh Nhu. "Nhu, Richardson said, was a great nationalist. When I mentioned some of Nhu's anti-American remarks and the resentment many anti-Communists felt toward him, Richardson said that the anti-Americanism was simply a product of Nhu's nationalism. He was a proud Asian, but he was also for us; more important, he was the one man who understood the strategic hamlet program. As for Mrs. Nhu, she too was a nationalist—perhaps a bit extreme occasionally, and sometimes a little emotional, but that was typical of women who entered politics—look at Mrs. Roosevelt."

To Halberstam, that sounded like such complete bullshit, he couldn't believe my father was sincere. Later he decided he must have been trying to manipulate him, to win him over or maybe just confuse him.

My father was too busy to worry about it. That fall, he spent a lot of time up in the Montagnard villages with Lou Conein or Colonel Le Quang Tung, a man often described in history books as one of the most hated and feared men in Saigon. He worked on a new national interrogation center—I hate to think what went on there—and got enthusiastic about a proposal from Nhu for a new guerilla program. He dispatched infiltration teams across the border into North Vietnam. From the regiments all the way to the joint chiefs of staff, he matched intelligence advisors to Vietnamese officers. Rufus Phillips told me that when they flew to strategic hamlets, he would talk passionately about the village leaders who had been beheaded and how much the poor Vietnamese needed help. Later, some of the central players said his work gave them all hope during this period. General

Earle Wheeler said his work on local intelligence stood "near the top in terms of progress achieved in 1962." Roger Hilsman singled out "Richardson of the CIA" as one of the men doing superb jobs in an uncertain situation.

But a few core problems kept recurring. The Montagnards were a perfect example. Stone-age tribes that dressed in penis gourds and carried spears, they lived in long houses built from a design that dated back thousands of years. For centuries the Vietnamese dismissed them as savages, ignoring and abusing them. When some of them finally rose in protest in 1958, Diem's army punished them by taking away their spears and crossbows. But the CIA realized how useful they could be and gave them back their weapons, turning them into eager warriors. By the time my father arrived, nearly two hundred villages had joined the program. Excited by their potential as a border patrol, he started establishing new camps and border surveillance units as fast as he could—within a year he had more than ten thousand tribal "strike force" members carrying out ambushes and reconnaissance patrols.

But the Vietnamese wouldn't cooperate. The province chiefs still treated the tribesmen with contempt. The Vietnamese Special Forces didn't want to fight alongside of them. And Diem and Nhu kept trying to confiscate their weapons and put limits on how far across the border they could wander. They didn't want thousands of armed tribesman loyal to the CIA.

The final snarl came from my father's efforts to implement Operation Switchback and turn the program over to the United States Army. The way the chain of command worked, this meant putting the tribesmen under Vietnamese leaders. But the Montagnards didn't want to fight alongside the Vietnamese, so they melted back to their villages or stopped fighting. Over and over, my father's Green Berets were forced to take command. Over and over, Operation Switchback got pushed back. It was the story of Vietnam in a nutshell.

One day at a Montagnard training camp up in the mountains, my father ran into another famous Vietnam journalist. Twenty-five then and just a few months out of the Army, Neil Sheehan found my father friendly and enthusiastic. But thirty years later one image stuck in his mind: on a hot day when everyone else was wearing fatigues or short sleeves, my father made his inspections dressed in a black business suit. "He was old school that way," Sheehan told me.

Finally we got all of our furniture and the house was ready for our first big party. It was New Year's Eve. All day long, the servants swept and mopped and bustled, polishing the marble floors with coconut husks.

That morning, my father had taken a small plane up to Da Nang to visit the training camp with Lou Conein and Colonel Patton. They were all planning to come back for the party.

When the phone rang, I was helping my mother make fish-house punch, pouring bottles of booze into a big crystal bowl. She had a bottle of wine in one hand and a bottle of cognac in the other. The maid came in and said it was urgent, very important, come right away.

It was a Vietnamese general. He said my father had been shot down and killed.

When my mother came back, my sister and I were stirring the punch. She said she had to go but would hurry right back as soon as she could. "Please be grown up and greet the guests," she said.

She jumped into the car and told the driver to take her to the airport.

Then everything was very quiet. Jennifer and I didn't know what to think. We concentrated on the adventure of doing the best we could until our mother got back. At some point, the guests started arriving, the Meryns and Rufus Phillips and lots of other people from the embassy plus Halberstam and Neil Sheehan and Malcolm Browne, a reporter for the Associated Press. My mother didn't want them to come but my father "insisted quite strongly," she remembered later.

On the way to the airport, my mother's driver fell in behind an ambulance that seemed to be headed in the same direction.

But she was expecting to meet the plane that would be carrying his body.

By the time she got to the terminal, she was so frantic she ran through the doors crying out, "Where is he? Where's my husband?"

Nobody knew what she was talking about. "I was told he was killed and you were bringing the body back to Saigon," she cried.

More blank looks.

She asked to use the phone. When she got through to my father's secretary at the embassy, she blurted out the same questions: "Where's John? What happened?"

His secretary gasped. "You weren't supposed to know about that!"

"Where is he? What happened?"

He was all right, the secretary said. He had a minor flesh wound. "I think they've taken him to a hospital."

"What hospital? I want to see him!"

The secretary told her to go home and wait. So she did. When she got home she found us bowing and curtsying to the guests like little angels, and went upstairs to freshen up—where she found my father shaving. "What are you doing here? You're supposed to be hurt!"

But he wasn't hurt, he said. The plane took off from an airstrip in the jungle, clipped the trees, and crashed. He walked away with nothing but a cut on his hand. The others were fine too.

"But you're supposed to be hurt!"

Sorry, he said. Although it was a bit tense waiting in the jungle to see which army would get to them first. Lou Conein and Colonel Patton took up positions but he was unconscious at first and didn't even have his gun. It was hidden in a shoe box in the closet, the bullets in a drawer.

So they went down to the party and joined the guests. To make the reporters feel welcome, my mother told them the story of the Buddhist exorcism. Later Patton came and told the story of the crash more vividly than my parents would ever tell it. They were about a hundred feet off the ground when the engine failed and the Turkish pilot brought them down in a mahogany forest, dragging the tail of the plane on the ground to slow them down and smashing the left wing against a mahogany tree. He had the longest seconds of his life waiting for the pilot to turn off the engine while gas poured out all over the place. Then they hopped out of the plane and ran about a hundred yards into the jungle, worried about Vietcong. My father looked at him and said, "George, I think I saw a sawed-off shotgun in the aircraft before we took off. Do you have it?"

"No," Patton said.

"I think you ought to get it."

Soaked in gasoline and knowing the plane could explode in flames at any second, Patton ran to the wreck and found the shotgun.

"I'm glad you found it," my father said. "Didn't I see a box of shells in there?"

"Yes sir, you did find a box of shells."

"Do you have that?"

"No sir, I don't have it."

"Well, I think we ought to have it."

When Patton finished the story, he laughed. "Like my father used to say, 'I guess I'm being saved for bigger and better things!'"

The next day, my sister sat down and took out her pink Barbie Diary (Property of Jennifer Lynn Richardson, age 10, grade 5). Opening the padlock with the little brass key, she began to write:

> Dear Diary—yesterday my father was in a plane crash in communist territory. He was with Mr. Conein and Col. Patton. Daddy was hurt on his right hand and passed out after walking away from the plane. Only Daddy was hurt. He only had a swollen hand. We had a New Year's Eve party. I stayed up till 3 o'clock. It was fun.

Two days later, in a battle just thirty-five miles from Saigon, the Vietcong stood their ground and proved once and for all that they could fight a "real" war, killing three American advisors and shooting down five American helicopters. This was the battle of Ap Bac, an important turning point in the war and especially in the media war. In his battle report in the *Washington Post*, for example, Sheehan said the Vietnamese soldiers refused direct orders to advance and an American Army captain was killed while pleading with them to attack. Halberstam and other reporters wrote similar stories, all accurately describing the South Vietnamese army's reluctance to fight and the impossible position of the American advisors who were placed in the line of fire but allowed neither to fight nor give orders. But admitting failure didn't fit with either military egos or with Kennedy's instructions to build Vietnamese moral, so General Harkins put out an absurd statement that the disaster was really a victory because they "took the objective," though everyone in Vietnam knew that was a meaningless statement out of a different kind of war. What they took was a few bloody rice paddies that would be back in Vietcong hands by nighttime. With that, any lingering respect the reporters had for the American officials in Vietnam went off a cliff.

The very next day, my father's programs took a parallel hit when hundreds of Vietcong troops attacked a Montagnard village named Plei Mrong, killing more than sixty tribesmen. Soon afterwards, drawing on my father's reports, CIA analysts back in Washington issued a secret memo describing the war as a "slowly escalating stalemate," an assessment far closer to the skepticism of the American reporters in Vietnam than to General Harkins's optimism.

A month later, the same analysts sent CIA director John McCone the first draft of another gloomy summary, listing many of the same failures the reporters saw at Ap Bac, including "lack of aggressive and firm leadership at all levels of command, poor morale among the troops, lack of trust between peasant and soldier." It was accurate, even prophetic, but in the meeting that followed, at a conference room stuffed with intelligence experts, McCone suddenly launched into an attack. He told the analysts to start over again, to go to Vietnam and talk to "the people who know Vietnam best." Later, an official CIA study tried to get at his motivation, citing his friendships with senior military officers and his concern about the consequences of a Communist victory in South Vietnam. But my father may have played a small role in his decision. Along with Harkins, Nolting, Hilsman and Colby, he was one of McCone's "people who knew Vietnam best" and one of the people McCone instructed the analysts to interview. And despite the skeptical tone in his field reports—which "consistently warned of the deteriorating situation and the possible consequences," according to the CIA study—my father seems to have joined the other top Saigon officials in an attack on the gloomy draft. "In their view," the study concluded, "it was simply wrong in judging that the Viet Cong had not been badly hurt."

McCone accepted the final version of the intelligence analysis on April 17. This time it said what he wanted it to say. "We believe that Communist progress has been blunted and that the situation is improving."

As the CIA later concluded, "policy wish" and "operational enthusiasm" had overwhelmed accurate intelligence analysis.

In the Barbie diary, my sister described full days, with boys on the schoolbus and ballet lessons and diving lessons almost daily with a Major Craft—the swan dive, the jackknife, the fishhook. She cleaned my mother's paintbrushes. She saw Doris every day. Sometimes I played with them, trying to convince them my new Ken doll was actually Superman. Or I would ride my bike with Mercy, which meant circling the driveway while she watched every move I made.

One day we went on a boat ride with Doris and a kid named Mark Cini. We took off our shoes and splashed our feet off the edge, waved to lots of naked Vietnamese kids.

One time I stayed with Jennifer and Doris at the Meryn's house. When they teased me out of the room so they could put on makeup alone, I crawled in bed with Marga and Sam. Sam was a kindly distant man who lost his family in the Holocaust, so he enlisted in the Army and then the CIA. After the war, he got caught giving radio equipment to some Czechoslovakian spies and spent a month in a Soviet prison. Marga was a jolly woman who was always hugging me and calling me Chacko in her Austrian accent, always cooking up blintzes or crepes. She worked for my father too. Secretly, I sometimes wished I could stay with her.

Later that month someone tossed a grenade over a neighbor's wall while they were having a garden party just like the ones we often had, globe lights hanging in the trees and waiters in white coats and me and Jennifer circulating with silver trays of canapés. Four people were wounded, one killed. Now sidewalk cafés were closing and there was steel netting on windows, steel grills on the shops. You could hear bombs going off in the distance. Armed MPs rode with us on the schoolbus.

People started leaving then, just as they had in occupied Vienna during the squirrel-cage days. One of my mother's friends took her kids. So did the wife of the press officer, John Mecklin. My mother scorned them for abandoning their husbands.

We got out of town too, but only for a week, flying up to a beach resort called Na Trang with Doris and Marga. We went out on a fishing boat and I dove into the water to help catch lobsters, which the sailors cooked right on the deck and served with the stinky fish sauce called *nuoc mam*. Some people worried that the Vietcong were in Na Trang too but my mother and Marga just joked about it—maybe the Vietcong were on holiday too!

My father and Sam came for the weekend, landing on the hotel's front lawn in a helicopter. But the big drama came when we went to watch giant sea turtles lay their eggs on the beach and saw some Vietnamese boys taking the turtle eggs, which wouldn't have been so bad except they were also flipping the turtles over for a laugh. My sister and I got so mad at the terrible cruelty and injustice we ran along the beach turning the turtles back over.

Over the next few months, the war escalated and doubts grew. Senator Mike Mansfield released a gloomy and influential report saying that the Army's official optimism was a fantasy based on the strategic hamlet program, which assumed that the peasants wanted to huddle in little armed villages an inconvenient distance from their fields. By this time Averell Harriman was already using his power in the State Department to argue against military involvement, so he grabbed the opening presented by the Mansfield Report to send new instructions to Ambassador Nolting. It was time to start talking to the dissident generals and politicians who were opposed to Diem, he said. Mansfield and Harriman may have been overreacting. I've read accounts that say the strategic hamlet program was a failure from the beginning and others that say it was just growing too fast. But their gloomy proclamations sent Diem into a defensive posture that made things worse. When Nolting went to the palace that April with a routine document authorizing another year's funding for the strategic hamlets, he refused to sign. There were too many Americans involved, he said. They were beginning to undermine the authority of his government.

Nolting reacted with alarm, asking Diem to consider carefully. Did he really want to cut himself off from American help and advice at this critical time?

But Diem held his ground. A country had to run its own affairs, he said.

It's a detail that doesn't fit into the cliché of a puppet government desperate for American help. Immediately, my father sent his agents out to find out how much truth there was behind it. A few days later he went to the palace to question Nhu face-to-face. Soon afterward, he sent a secret cable to Washington. According to Nhu, he said, this truly wasn't some kind of diplomatic chess move. Diem was completely serious. He had spent too many years resisting foreign interference to let Americans tell him what to do now. So they were building up evidence to make a formal case that the U.S. presence was too large and unmanageable. Demonstrating a cultural sensitivity that gave his report an extra ring of truth, my father even singled out his own Green Beret troops as "the main irritant" because they didn't always work well with local authorities.

That spring, as the Vietcong took over the Mekong Delta, my sister and I went to the movies at the American theater—*The Roy Rogers Show*,

Journey to the Center of the Earth, The Lady Is a Tramp. When we got home she made notes in her Barbie diary about what was really on her mind:

> I saw Patrisha sitting with Greg who I love.
> I sat next to Patrisha and Greg who I love.

But I wasn't with her the day the bomb went off. According to news accounts, there were about two hundred and fifty people at the theater that day, mostly American kids accompanied by their maids or mothers. Jennifer was with Mercy but really huddling with her girlfriends and ignoring Mercy as much as possible, just as her girlfriends were trying to ignore their maids (it was so dumb being followed around by maids everywhere you went). She remembers a big explosion from the bathroom or maybe from behind the screen and the scramble to get out of the theater. At some point, Mercy grabbed her hand and dragged her to the nearest cyclo. "I remember driving off, really, really scared."

It was our first glimpse of the trouble to come.

23

•

BUDDHIST TROUBLE

In May 1963, South Vietnam began to fall apart. It started during the celebrations for Buddha's birthday in the city of Hué. Thousands of people massed in the streets, many of them upset about a recent edict against flying Buddhist flags. With fiery speeches and a march on a radio station, a group of radical Buddhists was trying to stir this dispute into a mass protest against Diem. What happened that night is still disputed to this day, a *Rashomon* that splits along partisan lines. Either the police fired on the crowd or the Vietcong set off plastic explosives. Either the bodies were examined by a "distinguished physician" who confirmed the Buddhist story or the bodies were examined by a prominent Buddhist doctor who disputed the Buddhist story. What is certain is that Diem reacted badly, lecturing the Buddhists in his usual aloof way—religious freedom was already guaranteed in the constitution, didn't they know that? How dare they protest and issue ultimatums?

Diem's harsh response turned a local dispute into a focal point for urban dissent, bringing Buddhists and students into the streets by the thousands. Over the next days and weeks, they gathered in large crowds in Hué and Saigon, praying on their knees while their leaders insulted Diem on loudspeakers. Then the police attacked while the TV cameras caught the action, sparking new protests that sparked new attacks. A decade earlier, Diem might have been able to get away with it, but technology and the values of a new generation had changed the rules. David Halberstam's metaphors revealed as much: the protests took on the flavor of "civil rights rallies in Southern Negro churches," and Diem's police waded into the crowds just like Bull Connor's police did in Birmingham, enforcing segregation with police dogs and fire hoses. It all seemed part of the same thing, a "conflict between generations," a protest against authority, a clash between the forces of stiff righteousness and the young who were yearning to breathe free.

One day, my father ran into Neil Sheehan, a reporter he liked and respected. As he remembered the exchange later, he told Sheehan that the CIA had carefully studied the alternatives to Diem and couldn't see a single plausible candidate. "Do you want us to take a flying leap in the dark?"

"Yes," Sheehan said.

Behind the scenes, the United States kept pushing Diem to do something to calm the protestors—to make a statement or make concessions or just make nice. After Ambassador Nolting left on one of history's most poorly timed vacations, Deputy Ambassador Bill Truehart dropped the gentle approach and started going to the palace almost every day to apply "direct, relentless, table-hammering pressure such as the United States had seldom before attempted with a sovereign, friendly government," as one observer put it. But Diem held his ground.

During this period, an odd thing happened. Nhu summoned my father to his office along with Truehart and General Harkins, telling them he had just received a report from a confidential source that the Vietcong were about to pull six battalions back across the border. Should they take advantage of the moment and attack?

Harkins and Truehart didn't know what to think. Was the report accurate? Was Nhu trying to trick them into a pre-emptive attack? Was he trying to reassure them, showing that he was still ready to fight? While they hemmed and hawed, my father took the opportunity to push one of his pet projects, permission to expand the Montagnard border scout program. "It would be helpful if President Diem would authorize deeper cross-border intelligence operations," he said. Buried in the suggestion was a subtle dig about the restraints that left them unable to judge the truth of Nhu's report, his own bit of Oriental diplomacy.

As the protests continued, my father couldn't sleep at night. He tried not to let my sister and me notice, but my mother would see how restless and worried he was, tossing and turning and getting up and then lying back down, constantly frustrated over the charges that Diem was repressing Buddhists. "It just isn't *true*," he would say. "It's not *true*."

He sent cables to Washington trying to explain the nuances. Although there were some examples of favoritism toward Catholics, there were no formal restrictions on Buddhism. And while he couldn't find any evidence that the Vietcong were actually behind it, as Diem's partisans claimed, it seemed clear that some of the Buddhist leaders were using the protests to try to bring down the government.

In retrospect, he was right. Most scholars now agree that Vietnam actually had a lot of religious freedom, especially when compared to other countries in the developing world. But at the time, with police clubbing Buddhists in the streets, some found his reports questionable or even biased.

Then a Buddhist monk named Thich Quang Duc sat down on a busy street and set himself on fire. The press was there, called in advance. As another priest repeated over and over into a microphone, "A Buddhist priest burns himself to death, a Buddhist priest becomes a martyr," Malcolm Browne snapped a picture.

The next day, Browne's picture appeared on newspaper front pages across the world.

Maybe it was all the agitation in the streets, but during this period I got so restless and mopey my mother decided I needed a hobby. So she enrolled me in a woodworking class at a Catholic school on the outskirts of town. I took a few classes and liked it so much, I started imagining myself a famous woodworker. But the embassy found out and told my mother it was too risky. The school was too remote, too vulnerable to Vietcong attacks. They didn't want us traveling that far anymore.

So we went to the Cirque Sportif sports club instead, trying to while the days away with tennis and swimming. We never went alone. Usually Mercy took us. Then came the day when Jennifer and I finished up and got dressed and there was no embassy car to pick us up, so we got a taxi at the taxi stand. As usual, Mercy took her place guarding the door, blocking it with her body like that mythical three-headed dog. When we got to our street she told the driver to turn and he kept on going. She told him he missed our street and he ignored her and just kept driving. That's when she started to scream at him. At that point, Jennifer and I were more embarrassed than alarmed. Mercy was always getting fiery about something, flashing her black eyes at us and slapping her flip-flop against her thigh. But the driver just kept racing along and ignoring her and we began to feel uneasy. Now he was turning onto the road that led to the airport—where was he going?

He was trying to kidnap us.

Then Mercy pulled out her umbrella and swung it hard against his head, startling him so much he jerked the wheel into the curb. The cab stalled out and Mercy popped open the door, pushing us out.

Then the taxi driver roared off, leaving us standing in the dust in a daze of surprise and alarm. We never knew if he was a Vietcong agent trying to snatch the CIA chief's kids or just some random bozo swept up by an impulse for a big score.

A few weeks later, my model airplane got stuck on a ledge just a few tantalizing feet higher than a kid with a broom could reach. So I climbed up the garden wall and stood up straight and stretched to the tips of my fingers, and still the ledge was a couple of inches too far. So I jumped and caught the edge, scrambled up and inched along the ledge to the plane. Sailing it down to the yard, I felt like a hero.

Then things got tricky:

19. I'm hanging from my fingers. The wall is underneath me, four inches beneath my feet. I picture myself landing on the wall like a circus tumbler, catching my balance. Then I let go.

20. I'm in a yellow hallway, lying on a bed with wheels. And there's my mother looking very worried. She tells me that my arm is broken and my father is on the way and I decide that I'm not going to cry no matter how much it hurts—I don't want my father to see me cry. Then she asks me what I was doing and for some reason, I tell her I was chasing a lizard.

On June 25, my father met with Nhu and tried to persuade him to persuade Diem to compromise with the Buddhists. In an excellent little book called *Kennedy in Vietnam*, a historian named William J. Rust said that Nhu responded with a crazy fit of anger, lashing out at the Buddhists and even at his brother. Afterwards my father wrote a secret report saying that a state of emotional shock had put Nhu in a "dangerous" frame of mind, Rust said. The report ended with an alarming bit of speculation:

"It is possible that Nhu would lead efforts against Diem should he feel that a point of drastic deterioration occurred."

If Nhu was so unhinged that he was turning against his brother and telling a CIA chief about it, the situation had become truly drastic. But nearly forty years later, thumbing through my mother's papers, I discover a manila envelope with eight neatly typed letters inside. They are written on embassy letterhead, signed by my father. The first is dated on the same day as this alarming meeting, but gives no hint of any trouble at all, sticking to a weirdly detailed description of his flight from the Philippines a few days earlier. "I caught a MATS flight at 10:00 A.M. and arrived in Da Nang between 4:30 and 5:00 P.M. From Da Nang, I was able to board immediately a C-46 and found myself back in Saigon by 5:30 or 6:00 P.M. Wednesday afternoon." Skimming down, I see that my mother and sister and I were going back to the States for a vacation. Apparently, he accompanied us as far as Manila. Now he was trying to soothe us all with the format that had become most familiar to him, turning a family letter as dense and formal as an intelligence report:

> At the moment, I would be inclined to feel that we should be able to avoid further crisis with respect to the Buddhist issue. In other words, I think we are gradually coming out of the woods. However, these matters are unpredictable and I may find myself proved wrong two or three weeks hence.

On he went, telling us that Diem's proposed peace agreement with the Buddhists—along with promises for general reforms, he agreed to compensate the victims of the Hué violence and punish the people responsible—was already running into skepticism. Some people were sure he was going to drag his feet. And there was quite a bit of local controversy over the way the crisis was being covered in the press.

On the whole, my impression is that the local American reporters have given a fair and judicious account of developments. However, I gather that American editorial writers back home took off in a rather violent way against the administration here.

Trailing off, he mentioned a lovely dinner with Bill and Phoebe Truehart and put in a few lines about his last golf game with Colonel Tung, a pleasant eighteen holes on a rainy Sunday—he shot a ninety-four. And please tell little Jocko he tried to call Manila but couldn't get a decent connection.

At any rate, I did try to get through to him to wish him goodbye. Many people have been asking about him and have wanted to know how his broken arm was progressing.

He closed with "all my love."

Just three days later, he sent a much different report to Washington, a long and discouraging cable that described trouble on the border with Laos and continuing strife with the Buddhists. Diem was convinced he was in a struggle for survival, he said. It was very unlikely that he'd make any significant political reforms and equally unlikely that he'd replace Nhu or Madame Nhu under pressure. If we pushed the issue, he'd consider it a personal insult. As for bringing opposition politicians into his government, this would only force him to encourage enemies who seemed to have no intention of compromising with him anyway.

But he might still ride out this storm as he had others.

Four days later, the *Times of Vietnam* enraged the Buddhists with a piece suggesting that Thich Quang Duc was on drugs when he burned himself, a charge especially provocative since everyone knew that Madame Nhu was the power behind the newspaper. Two days after that, Halberstam shocked Diem with a front-page article about dissension in the Vietnamese military that seemed to many readers, especially Vietnamese readers caught in the tinderbox of local politics, like an attempt to promote a coup. Shortly afterward, a group of Diem's police attacked a reporter named Peter Arnett, possibly mistaking him for Halberstam, and Halberstam jumped into the melee and helped Arnett get loose and then led a small group of outraged reporters to the American embassy to demand a formal diplomatic complaint. Sometime during that same crazy day, probably not long after Halberstam stormed back out of the embassy in a fury bigger than the one that brought him, my father sat down and wrote us another astonishingly dispassionate letter:

> I have just received your letter of June 27th from Waikiki and consequently have learned of Jocko's hospital experience there. As you can imagine, I was extremely sorry to hear that his arm had to be reset.

He was also sorry to miss my sister's eleventh birthday, he said.

> I bought her a very nice dress which I will hold as a surprise for her on her return to Saigon. I feel sure that she will be extremely pleased with it. It will help her to look more grown up and is really very attractive.

Then he turned to business, giving us the news that the president had replaced Nolting with a new ambassador, Henry Cabot Lodge. He didn't say anything about how shocked everyone was that Kennedy would fire Nolting during his vacation, how worried they were about what this would mean for the overall Vietnam policy. He just said he would miss the Noltings, that working with them was "one of the most pleasant and constructive experiences" of his career. And of course Bill Trueheart continued to do a magnificent job. But as far as the Buddhist crisis went, things seemed to be quiet at the moment.

On another positive note, he said, he flew up to Na Trang for the final turnover of the Green Berets to the Army. That left him with just one guerilla program still under his control, the mountain scouts patrolling the border. Given the resistance the Montagnards had to Vietnamese supervision and his old strategic fear about the permeable border, he couldn't quite give that one up. And he'd been able to get some golf in too, running scores of ninety-four, ninety-six and ninety-eight.

It was good to have a break in the weekly routine of work.

And things were fine at the house too.

> Mercy is getting along fine and Frisky and Meeno are in good shape. I now have my weight down to 166 pounds and

intend to aim at a target weight of 160. This is going gradually and I feel myself in good health.

Again, he closed with "all my love."

The next day, in a cable to Washington, he took a very different position. Diem's enemies had begun plotting coups, he said. There was even
talk of assassination. Meanwhile the Buddhists seemed determined to agi-
tate until the government collapsed.

All of this put Diem in a tough spot. If he cracked down, the protests
would increase. If he tried to make peace, his enemies might see it as a
weakness. If he did nothing, the situation would drift into greater danger.

A week later, he sat down and wrote us another letter.

I assume that by now you have traveled over the western
part of the United States, have seen the Grand Canyon and
the Painted Desert, and have arrived in Detroit. This must
have been a grand experience, and I shall look forward to your
telling me about your adventures upon your return to Saigon.

As before, he included a surprising amount of detail about his work.
Nolting had returned to pack up and prepare the embassy for the turnover,
they were making steady if undramatic advances against the Vietcong,
Diem was handling the Buddhist situation with a fair amount of restraint.
But things were still somewhat tense.

There are rumors about coups d'état and possibly new acts
of self-immolation on the part of misled zealots. There could
be a sudden reversal to crisis at any time depending primarily
on whether some bonze sets himself afire or eviscerates himself.

The next morning, as the Buddhists started protesting right in front of
Nolting's house, word spread that another monk would burn himself soon.
Two more tense weeks followed. After pleading with Diem in a series of
meetings, Nolting finally got him to promise to try to placate the protesters.
Then Nhu showed up at Nolting's goodbye party and went into a long rambling monologue that ended with a defiant flourish: "If the Buddhists wish
to have another barbecue, I will be glad to supply the gasoline and a match."

My father must have been there. I doubt he got much sleep that night.

But two days later, he sat down and wrote to us again, saying that he was delighted by our stories about the World's Biggest Copper Mine and swimming in the Great Salt Lake and that he was still playing golf, both Saturday and Sunday for the last two weeks. And he just had lunch with Marguerite Higgins, the famous Korean War reporter. On Friday, he was going to have her over to the house for dinner with General Stillwell and some fellows from the strategic hamlet program.

Trying to stay positive, he didn't mention the Kennedy advisor who had arrived to investigate the poisonous split with the American reporters. Instead he said that Higgins—who was edited by Si Freidin, an old friend from Vienna—was one of many new reporters who had arrived, which should help the situation.

> Our impression is that even the local U.S. newspapermen are becoming a little disenchanted with some of the Madison Avenue techniques which the Buddhists have carried a little too far. They found the stage managing too slick and the proposed burning revolting.

But it was just another glint of silver in a thunderhead.

24

•

KILL TUNG FIRST

As August began, another Buddhist burned himself and Madame Nhu told a reporter that next time, she'd clap her hands. My father sat in his office and wrote a long cable with an optimistic lead—"Nothing occurred to change our basic belief that the VC will eventually be defeated"—followed by a dark little army of hedges and asides. Military programs to harass the Vietcong were "difficult to assess." The enemy had been hurt everywhere "except the Delta." The strategic hamlet program had "failed to live up to expectations." Success hinged on whether "the Buddhist crisis is resolved shortly in a reasonably satisfactory manner." The problem of patrolling those long borders "is a staggering one."

On one of these hot troubled nights, he had dinner on the roof of the Hotel Caravelle with Maggie Higgins and a British counterinsurgency expert named Robert Thompson. As Higgins told the story, she asked them what would happen if someone pulled off a coup against Diem.

"It would set the war back twelve months—maybe forever," Thompson said.

My father nodded agreement.

She asked for their reasoning and they gave an explanation few historians mention. There were forty-one provinces in South Vietnam and in each one, the generals would replace the chief with a man of their own. That was just the way things were done, the routine change-of-government purge. But it was hard to find good administrators, harder still to get them up and running. Diem had been working on this for nine years. And all that work would be destroyed overnight. Chaos would follow.

"Military juntas are notoriously unstable," my father added.

Higgins went on to a five-hour interview with Diem, who told her he was under attack and fighting for his life. So why did the American newspapers call his government a "Catholic regime"? Was the Kennedy government a Catholic regime? And why did they attack him for leaning on his brother? Didn't Kennedy also lean on his brother? And why press him on civil rights when he was in the middle of a war? Didn't the United States also have civil rights troubles?

Higgins asked Diem if he really thought the United States was plotting against him. He chose his words carefully. "I do not think Ambassador Nolting is plotting against me. I do not think Richardson is plotting against me."

But he knew that others were preparing the way.

Then Madame Nhu exploded across the newspapers of the world. "What have these so-called 'Buddhist leaders' done?" she asked one reporter. "All they have done is barbecue a bonze." In a *Time* magazine cover story, she even sank her claws in Diem. "He would like to conciliate as the Americans desire, smooth, no bloodshed, everyone shaking hands." The combination of her glossy black hair and tight dresses with her militant feminism and cruel tongue seemed to drive reporters and editors into furies. Like a scene out of Shakespeare—Dragon Lady Macbeth—it had the same effect on Diem's opponents in Vietnam. That week, another monk burned himself.

In the thick of this crisis, my father wrote us four letters in four days.

> Things are going along here much the same with no dramatic changes. Madame Nhu's public statements have, as you can imagine, been greatly irritating American and foreign

sensibilities, but we still feel the Buddhist crisis is gradually running out of steam.

Yesterday he played with General Stillwell and Colonel Tung, he said, shooting a ninety-five and "finally beginning to get some wrist snap into the game."

The next day, he responded to a letter from my sister, saying that he was glad she was spending time in the outdoors—some of his happiest days were spent in the mountains and woods. And he was very pleased she took the time to write him such a sweet letter and wanted her to know that he loved and missed her very much.

A day later, he wrote a letter just to me.

> I would never have believed it but, as far as letter writing is concerned, you have turned out to be the most silent member of the family. I know that you have a good excuse in view of the fact that your writing arm is probably still laid up. However, if you are still on speaking terms with your mother or sister, I should assume that you could dictate several pages in about five minutes of fast talk.

He went on for two more paragraphs, picturing me swimming and basking in the sun and having a wonderful time. Hopefully I would manage to get in a little trout fishing.

That series of letters broke off when another Buddhist committed suicide by fire, this time a nun. The next bit of writing my father did appeared as an attachment to a letter from Bill Colby to Dick Helms. A study of the various coup plots, it read like an entry in the *Meditations*.

> While perhaps conceding Nhu's competence to hold high office, in terms of experience, organizational capability, and as the driving force behind the strategic hamlet program, etc., there exists considerable opposition to him among the educated and articulate elements of the population, including the military. Unquestionably, his greatest liability is Madame Nhu, toward whom these same elements express an intense and indeed very personal hostility on the grounds that she is

vicious, meddlesome, neurotic, or worse. Whether this opposition to Nhu and his wife is based on cold logic or on supercharged emotion is immaterial, it is important because it exists.

He didn't hold back the ugly details, mentioning a dinner party when Nhu "gradually worked himself into a highly emotional state of mind" and began verbally attacking Diem. There was also the bizarre interview Nhu gave two reporters from *Time/Life* when he said the government would have to be overthrown, repeating himself several times and ending with the Latin phrase *Carthago delenda est*—Carthage must be destroyed. Either Nhu was in a fatalistic mood or he was threatening a coup of his own. He followed up with some suggested contingency plans and a detailed analysis of all the military forces arrayed around Saigon, from the six battalions of the Airborne Brigade to the four Marine battalions to two hundred Ranger companies and the twenty-five thousand Presidential Guards "heavily armed with ten M-24 tanks, six M-113 armored personnel carriers, six M-114 armored personnel carriers, vehicles with mounted quad 50s, recoilless rifles, bazookas." Then he gave the same meticulous treatment to the major personalities, from the general known as "Big" Minh ("probably has more prestige with other army officers than any other single army officer") to a relatively minor player like Major Nguyen Van Ba, head of the Second Armor Squadron ("his loyalty to the regime is unknown"). But none of them was strong enough to succeed Diem, and there was a good chance "coup will follow coup in an increasingly anarchic situation."

Later Kennedy and Lodge would claim to be shocked by the violence they unleashed. But the record shows that my father detailed the most ominous possibilities:

> Assassination may be an integral part of projected coups or may be done in hope that something better will somehow emerge from the resulting chaotic situation . . . Nhu and his wife would be fortunate to escape with their lives. . . .

In the same uninflected tone, he mentioned that most of the coup plots include plans for "eliminating Tung"—his regular golf partner.

25

PAGODA ATTACKS

On a quiet and pleasant night toward the end of August, someone cut the phone lines to the American Embassy. Busloads of South Vietnamese soldiers rumbled through the streets to the Xa Loi Pagoda, an important Saigon temple with three ornate peaked roofs surrounded by high walls and a spacious courtyard. This was the heart of the Buddhist unrest, the refuge for the leaders and the staging area for the protests. Sheehan and Halberstam arrived right behind the soldiers and reported what happened, the shattered glass and pistol shots and screams that mixed with the droning clang of a temple gong. For two hours the gong tolled and the soldiers dragged out monks and nuns, arresting some four hundred in all. The same thing happened in other cities across Vietnam.

There was one particularly chilling sight that night, Halberstam recalled in *Quagmire*—a squad of Colonel Tung's Special Forces carrying submachine guns. "As they pranced into the pagoda, looking something like a smart football team coming up to the line of scrimmage, the mark of American instruction was all over them."

The next morning, Halberstam headed over to the embassy to get the official reaction and pursue a hot rumor passed along by "responsible Vietnamese." Did the CIA know about the raids in advance? Did the CIA endorse them? After all, everyone knew the CIA trained those Special Forces troops. Everyone knew Richardson had a close relationship with Tung. And everyone knew that he had a close relationship with Nhu. And everyone knew that Nhu was the man behind the raids.

In Halberstam's version of what happened next, my father asked to see him in his office. "That morning Richardson was a tired and shaken man," he recalled. "He refuted the rumor immediately. 'It's not true,' he said. 'We just didn't know. We just didn't *know*, I can assure you.'"

In my father's version, they met by chance in a hallway and Halberstam asked if he'd known in advance about the raid. "It was a no-win question for me, but I told him the truth, namely that we hadn't. His subsequent story correctly stated that the CIA had been caught once again unawares."

By then, South Vietnamese police and army troops had cordoned off the pagoda and set checkpoints up to guard all the roads in and out of the city. Diem had closed the airport and put the press under military censorship. My father spent the day monitoring the situation, sending cables back to Washington. More and more students were joining the Buddhist protesters in the streets. Diem's former spy chief was plotting a coup. Diem had six battalions on alert around Saigon in case of trouble.

And Halberstam worked on his story. Soon a pattern began to develop, he wrote later.

> I met an American intelligence friend and asked him who was responsible. "Nhu and Colonel Tung," he said, and then he listed the units which had been involved. "It didn't have a damn thing to do with the army," he said.
>
> "Why didn't you know?" I asked.
>
> "We should have known," he said. "We had every damn warning you can get. We kept telling *them*"—meaning his CIA superiors—"that Tung had his special forces in town and that something like this could happen. We could have been spared this. We could have headed it off."

The next night, Halberstam had drinks with two CIA men and found them "exceptionally bitter" and "not happy with Richardson."

Convinced he had the inside scoop, Halberstam filed a story that blamed the raids entirely on Nhu, saying he planned the whole thing without the knowledge of the army. This has become the standard version in many histories, despite a counterstory that casts the usual Vietnam shadows. Well before the raids, for example, the South Vietnamese generals were pushing Diem to take action against the Buddhists. The day before, they voted unanimously to impose martial law and remove dissident Buddhists from Xa Loi Pagoda. Their troops probably participated as well, as an Australian scholar named Anne Blair flatly stated after a thorough study of the historical record: "The evidence is quite conclusive that Army of the Republic of Vietnam troops acting alone conducted the assaults in Hué." Also, General Tran Van Don said in his autobiography that Nhu told him and other top army commanders about the raids themselves at least a few hours in advance, although the army "as a whole" still didn't know. Bits of this alternate version reached the *Times*, and another reporter named Tad Szulc turned in a story saying it was actually the generals who pushed Diem into ordering the crackdown. On August 22, the *Times* published both stories on the front page under an extremely cautious headline:

TWO VERSIONS OF THE CRISIS IN VIETNAM
ONE LAYS PLOT TO NHU, OTHER TO ARMY

In a drizzling rain, Henry Cabot Lodge arrived at the Saigon airport to take charge of the embassy. Tall and handsome, a former senator and ambassador descended from several of America's most prominent families, Lodge was a supremely self-assured man who was used to getting his way. At the airport, he signalled a change in the official attitude by giving a speech to the assembled reporters about the vital role of the press in American democracy. The next day, he made an even more dramatic gesture by visiting two Buddhist protesters who had taken asylum at the embassy. Behind the scenes, he sent another disquieting signal by skipping the traditional arrival meeting with the "Country Team," directly snubbing my father and the other senior embassy officals. Hiding their concern, they waited to see what would happen next.

By that time, the American press was flush with outrage over the pagoda raids. In the *Times*, Nhu was "a kind of Oriental Richelieu," in *Newsweek* "an arrogant, anti-American intellectual." All over the country, publications carried the same anonymous quote from a returning journalist: "There has been no family like it since the Borgias." And Halberstam was already chasing down a new story about General Tran Van Don, who was telling friends he'd been given no advance knowledge of the pagoda raids.

As it happens, General Don invited Lou Conein to his office and told him the same story. As the chief of staff of the Vietnamese army, his word carried weight. But when Conein pressed him, he offered four important qualifications:

1. Diem made the final decisions.
2. Tung took his orders from Diem, not Nhu.
3. The generals came up with the idea for martial law themselves, without Nhu's encouragement or participation.
4. Afraid the Vietcong had infiltrated the pagoda, the generals even endorsed a plan to send some of the protesters back to their provinces, a plan that sounded so much like a raid on the pagoda that when they brought it up with Diem he asked them not to hurt any of the Buddhists.

Unreported at the time and still overlooked in most historical accounts, these qualifications raised important questions about the rush to blame everything on Nhu. They were the main bit of intelligence behind item number one on President's Kennedy's Intelligence Checklist that weekend: "We cannot determine as yet who is calling the shots in Saigon—Diem or the Nhus."

But things were moving too fast for nuance. The next day, Halberstam's story on General Don hit the front page of the *Times* and Lodge sent two cables, the first quoting a Vietnamese official who told him Vietnam would slip into chaos unless they got rid of Nhu, the second reporting a story that Nhu actually tricked the generals into martial law. And back in Washington, four of Kennedy's top advisors—Averell Harriman, Roger Hilsman, George Ball, and Michael Forrestal—went to work on the pivotal and infamous document that would enter history as the Cable of

August 24. "Eyes only Ambassador Lodge," it began. "No further distribution."

> U.S. Government cannot tolerate situation in which power
> lies in Nhu's hands. Diem must be given chance to rid himself
> of Nhu and his coterie and replace them with best military
> and political personalities available. If, in spite of all your
> efforts, Diem remains obdurate and refuses, then we must face
> the possibility that Diem himself cannot be preserved.

Then came the sentence my father would remember for the rest of
his life:

> Therefore, unless you in consultation with Harkins per-
> ceive overriding objections you are authorized to proceed
> along the following lines …

From there, the cable outlined a series of "prompt dramatic actions," starting with a demand for immediate concessions to the Buddhists and ending
with signals to the Vietnamese generals of U.S. support for their coup efforts.
There are still many disputes over who saw the cable before it went out, over
who signed off and what they were told and whether Kennedy even read it.
Later General Maxwell Taylor denounced the whole thing as "an egregious
end-run" by Harriman and his team. What is certain is that it hit the CIA
with a shock. When Colby saw the cable the next morning, he immediately
called CIA director John McCone, who was on vacation in California, and
discovered that nobody had contacted him. McCone said to rush the cable
out in person so Colby grabbed the next White House jet. Later he remembered that when McCone finally read the cable, he was furious, outwardly
calm but exceptionally icy. American policy to Vietnam had been changed
behind his back without so much as a consultation. They drove straight to the
airport and flew to Washington for a showdown at Kennedy's morning
meeting—but only after sending instructions to my father that under the circumstances, the CIA had to "fully accept directives of policy makers and seek
ways to accomplish objectives they seek."

By that time, the Cable of August 24th had reached Saigon. Lodge
immediately shot a cable back to Washington saying that he thought the
chances of Diem agreeing to dump Nhu were "virtually nil," that might

even alert Nhu and give him a chance to make some last-ditch effort to save himself. Better to go straight to the generals, telling them the choice was theirs—keep Diem and get rid of Nhu, or get rid of both of them.

From the State Department, without White House input, George Ball replied immediately: "Agree to modification proposal."

Now they were at the brink of a coup.

I'm not sure if my father saw the cable that day. In later years, Lou Conein told a colorful story about getting a copy from General Harkins and rushing it over to my father, only to hand it to him just as Lodge happened to walk into the room—and Lodge exploded in rage, cutting the old man out of the embassy cable traffic once and for all. But my father told me that never happened.

What the record shows is dramatic enough. That day at lunchtime, he got word that General Nguyen Khanh had approached a CIA officer named Al Spera and requested an immediate meeting. After pleading that everything he was about to say be kept a secret from Nhu, Khanh told Spera that Diem and Nhu were talking to North Vietnam about some kind of peace arrangement—an alarming bit of intelligence that echoed similar rumors from other sources. If the generals and the United States had one thing in common, it was hostility to any suggestion of peace with the Communists. So the generals were determined to stop taking orders from Diem and Nhu and wanted to know whether the United States would support them if they made a move.

As soon as he got Spera's report, my father sat down and put it into a long cable. The news was so explosive the CIA rushed his cable direct to the White House. A clerk even noted the exact moment it arrived: 10:41 A.M.

POINT OF NO RETURN

For some reason, wisps of the journey I took with my mother and sister that summer still stand out in my mind—the little ranch house in Whittier, where I learned how to sleep with a cast on my arm; the hotel in Las Vegas when Jennifer and I woke up to the sight of a thousand silver dollars stacked at the end of the bed, the basket of polished stones in some gift shop near the Grand Canyon.

By late August, we were staying in a cabin on a misty lake in Michigan and my sister and I were getting to know our maternal relatives. We barely knew we had a grandmother much less all these cousins. Then, my mother got a call from Washington. "If you want to get back to Vietnam," a friend warned, "you better get there in a hurry." The State Department was about to close the country to family members. But if you were already on your way, you could still get in.

We caught the first plane to New York, a blur of wet sidewalks and grand doorways. From there we went to Paris, where we stayed in a frumpy old hotel with steamy radiators and tall narrow windows, spending afternoons at a charming old house with books about an elephant named Babar, waiting to see what Dad wanted us to do next. He didn't want us to come back just yet, the situation was too uncertain. So the Paris CIA chief told us to go to Rome and the Rome CIA chief told us to go to Athens, where they gave us a car and a driver and an apartment and Mom killed time by arranging a meeting with our old nanny, reprising the family chestnuts about how I tagged along on her dates and begged her to wait for me to grow up. But by then she had stepped down from the clouds of my childhood into life—just another adult woman with a little daughter.

We went to Delphi and waited.

By now the whole world was in an uproar over Vietnam. The Burma *Guardian* was denouncing Diem's "mad war" against the Buddhists, the Manila papers were warning of a break in diplomatic relations, the French kept sneering about how the Americans hadn't learned a thing from the tragedy of Dien Bien Phu. In Saigon on the morning of August 26, the Voice of America officially pinned the blame for the pagoda raids on Nhu, warning that the U.S. was ready to cut its aid to South Vietnam if Diem didn't take action.

Shortly afterward, Lodge arrived at the embassy and summoned my father and General Harkins to his office. This seems to be when he finally showed them the Cable of August 24, a rather shocking delay that makes me think there might be some truth to Conein's story about my father getting an advance copy. But the way my father always told it, Lodge simply handed them the cable himself and waited for them to react. In silence, they read. Near the end, my father came to the pivotal line, the one that made the decision personal:

> ... unless you in consultation with Harkins perceive over-
> riding objections ...

He stopped to think. Hell yes, he had objections. He thought it was hasty and unwise, even an act of betrayal. For all his flaws, Diem was our ally in a time of war.

But were these objections *overriding*?

More to the point, did he have overriding objections to an order endorsed by the heads of the CIA, the State Department and the Armed

Services—a direct order from the President of the United States of America?

Was he so much smarter than everyone else?

Was he so much wiser?

Exactly what was said, no one has ever reported. But it seems that Harkins didn't raise too much of a fuss, though he also disagreed. So my father made his decision, which he later described in emphatic and revealing detail:

> I placed my station at the complete disposition of Ambassador Lodge and invited him to assign a senior foreign service officer of his own staff to participate in all station actions and decisions relating to the immediate matter at hand, which Ambassador Lodge did.

After leaving Lodge's office, my father sent off a private cable alerting McCone to the rush of new plans. As usual, his tone suggested a complete, almost bureaucratic detachment, but there was a hint of his true feelings in the unusually specific detail that began the cable: "Lodge made decision that American official hand should not show."

After the meeting, he summoned Conein and Spera and gave them their instructions:

Go and tell the generals that Washington is now open to a coup. Tell them everyone agrees that Nhu has to go. It's up to them whether or not to keep Diem. And please avoid bloodshed—or keep it to "an absolute minimum."

And so the August coup began. At lunchtime, Conein passed the message to a general named Tran Thien Khiem. Khiem said he would arrange an immediate meeting with "Big" Minh.

At the same time, Spera carried the message to General Khanh. But now Khanh was nervous. He wasn't ready to move just yet, he said. First they needed some assurances. The United States would have to promise to give the plotters and their families safe haven and financial support if the coup failed. And when Spera tried to goose him along by telling him that Conein was already meeting with Khiem, he blew up: Khiem can't be trusted! You're endangering the whole plan!

With that, he broke off the meeting and said he would never talk to anybody from the embassy again.

At five that afternoon, going through the official motions as if none of this was happening, Lodge went to the palace to present his diplomatic credentials to Diem. My father and General Harkins were supposed to go with him, but Lodge made one adjustment to account for the backstage plotting. Afraid that Diem might get wind of what they were up to and grab the opportunity to take all three of them hostage, he asked Harkins and my father to stay behind.

That leaves Lodge's memo of the meeting as the best record of what happened between the two men that day. Trying to forestall a Diem monologue and take control of the meeting, Lodge started right in with a prepared statement about the importance of public opinion and the need to placate the Buddhists. But Diem just told him that public opinion was wrong and kept explaining why it was wrong for the next two hours.

When he was finished, one thing was certain:

Henry Cabot Lodge did not like being lectured.

Back at the embassy, my father got the news about Spera's aborted meeting with General Khanh. Late into the night, trying to keep the coup moving before it collapsed, he and his men met with generals and spies. By morning he was able to report that most of the generals now said they were ready. Because of his control of the Vietnamese Special Forces and his role in the pagoda raids, Tung was a primary target and would "be destroyed together with his entire encampment as one of the first acts of the coup." Towards the end of the cable, my father added that he was complying with Lodge's orders to build trust with the opposition by getting the generals a complete inventory of the ordnance at Tung's camp. He also gave them assurances of safe havens and financial support, as instructed.

At that point, everyone began counting heads. Tung had almost eight thousand men. Most of the Army troops assigned to the capital were under the control of another Diem loyalist, General Ton That Dinh. The Saigon police seemed to lean toward Diem too. The forces of the dissident generals were stronger, but most of them were outside of the city limits, which could mean a bloody fight just to get into town. If Lodge cut off aid to Diem and told the U.S. military to help with communications and transport, that might make all the difference. After a long day of this, a day full of plotting and intrigue and round-the-clock meetings, my father sent the cable most often associated with his name:

> Situation here has reached point of no return. Saigon is
> armed camp. Current indications are that Ngo family have
> dug in for last ditch battle.

Things had changed drastically since the raid on the pagodas, he said. If Diem and Nhu won now, Vietnam would "stagger on to a final defeat at the hands of their own people and the VC." If the generals tried and failed, Diem and Nhu and the American public would all demand the same thing, a sharp reduction or even withdrawal of the American forces—and the Communists would win.

The generals seemed organized and committed and had a good chance of winning, he continued. But there was a chance of widespread fighting if Dinh and Tung weren't "neutralized" quickly.

And obviously it would be better if the coup went off without "apparent American assistance." Otherwise the generals would be vulnerable to charges that they were American puppets.

> We all understand that the effort must succeed and that
> whatever needs to be done on our part must be done. If this
> attempt by the generals does not take place or if it fails, we
> believe it no exaggeration to say that VN runs serious risk of
> being lost over the course of time.

When the cable arrived in Washington, Colby didn't know what to think. Later he told a historian named Francis Winters that it didn't "compute." Yes, he had given my father instructions to cooperate with Ambassador Lodge. But he didn't order him to suddenly change his mind and start believing in the coup. And this was a powerful and strongly worded cable that would have a big impact on Kennedy and his senior advisors, especially since it came from a man known to have reservations. As Winters said, it had "the authority of on-the-scene reality."

Colby had no choice. He took the cable to the White House, opening the noon meeting by reading it out loud.

Most of the senior Washington players were present that day—the Secretary of Defense, the Secretary of State, Averell Harriman, Roger Hilsman. The record of their conversation shows that they were shockingly out of touch with what was actually going on in Vietnam and still arguing about whether to endorse the coup. The Secretary of Defense said it was time to make a decision. Only Harriman seemed to realize how far things had progressed, insisting that we couldn't let the coup fail because Diem had "doublecrossed" us and we'd lose Vietnam if we stuck with him. But Kennedy seemed to think we weren't in too deep to bail out.

They adjourned without making any decisions.

In Saigon two hours later, our home phone rang. It was five in the morning. General Richard Weede wanted him in the Army's war room right away.

When he arrived, Conein was there too.

Weede told them that General Taylor had sent a cable to Harkins saying that Kennedy was having second thoughts. Could they stall?

So far as I know, what happened next is not in any history books. But many years later, my father told me that he turned to Lou Conein and asked him if he could stall without harming the coup. Conein said he could. But he'd have to send word in half an hour.

So he had half an hour to decide "whether to pull the trigger."

He would be taking a risk. Lodge would be furious if he found out. There would be repercussions. But Maxwell Taylor was a man he respected immensely, a man who commanded the 82nd Airborne during the Italian campaign and the 101st Airborne during the Normandy Invasion, who ran the United States Forces in Berlin when my father was in Vienna, who took over the Eighth Army for the final stages of the Korean War—a military legend who was now Chairman of the Joint Chiefs of Staff. And it would be an awful mistake to rush forward if the President really did have doubts. And maybe that brought his own doubts flooding back, and maybe he got into one of his stubborn moods. So he came back and told Conein to go ahead and meet with Minh—but just to listen, not to give the coup American support.

We could pull the trigger later.

There are several versions of what happened next. In the most dramatic, Lodge happened into my father's office just as Conein was giving his report. When he realized they hadn't pushed the coup, he became angry and demanded an explanation.

"Because of the 'second thoughts' cable," my father said.

And Lodge roared, "*Why wasn't I informed of this cable?*"

From that point on, he insisted that Conein report to him first.

The way my father told it, Conein gave his report and Lodge simply asked if the delay had in fact caused any damage to the coup effort—and Conein said that yes, it had. This came as a surprise to my father, who remembered Conein saying that it wouldn't harm the effort. But he remembered no anger from Lodge, no roaring or change in the chain of command.

One thing is certain, he had little time to think about it. That same day, from a "source of medium reliability," he had a tip that Diem was about to order the arrest of the generals he suspected of disloyalty. After conferring with Lodge, he sent an agent to pass the word to a Vietnamese officer via an Air Force colonel who lived next door. First the Air Force colonel tried his neighbor on the phone, but even though he knew the man was home there was no answer. So he went next door and knocked. Still no answer. Everyone seemed to be in hiding. Finally they made contact with a general who sent an aide to lead them through the lines of nervous troops. After passing on their message, they snuck home via a private gate and a back road.

Late that night, a source checked in with a hot report about a dinner attended by three generals and a colonel linked to Big Minh. One general said he didn't think a coup was a good idea. They had too much to lose. Another said that if someone killed Diem and Nhu, he'd support Big Minh. The rest agreed that if Diem were dead, they'd jump right in. From there the conversation turned to strategy and plans for future glory.

All day long, my father detailed these events in cables to Washington. Hints of his personal turmoil in the wake of the conflict with Lodge thread through them. "Hope HQS will understand that speculation on possible negative factors does not represent negative station attitude," he said in one. "We are bringing all capabilities we can to bear." Another ended with the news that he had in fact given the generals the complete list of the weapons controlled by Colonel Tung and even a helpful sketch of where they could be located.

> This is cited as another example of assurance re USG intentions since these plans were provided for purpose of attack on Col. Tung's training camp.

But another day went by and still no coup. That night Lodge wrote his famous complaint that getting the generals to move was like "pushing a piece of spaghetti."

The day after that, Harkins met with Khiem to give him his personal green light. But before he could get the words out, Khiem told him the coup was off, that Big Minh thought they didn't have enough troops to win. This put Harkins on the spot. Personally, he thought the coup was a mistake. So did he obey his orders and urge Khiem to reconsider? Or keep his mouth shut and wait to see what happened?

He kept his mouth shut.

At 2:39 in the morning, my father reported the news in a succinct cable:

> This particular coup is finished.

27

•

ON THE
ASSASSINATION LIST

As soon as the coup collapsed, the repercussions began. First Nhu stopped talking to my father. Then Madame Nhu's brother told an Australian journalist that the Americans behind the coup plans should fear for their lives, helpfully writing their names down on a list:

1. Richardson
2. Conein
3. Mecklin
4. Sheehan
5. Halberstam.

Around the same time, Madame Nhu told a reporter from *Der Spiegel* that she had just finished writing a story for the *Times of Vietnam*. Then the reporter heard her tell a staff member, "I want you to put Colonel Richardson's name in the story too."

I have a copy of that newspaper, its angry headline spanning the entire front page: CIA FINANCING PLANNED COUP D'ETAT. Although it didn't actually use my father's name, the story blamed "the CIA crowd" for whipping Buddhist agitators into a state of hysteria, for paying as much as twenty-four million dollars in bribes, for mobilizing student protesters and even for subsidizing articles by hostile foreign correspondents, all in a desperate attempt to undercut "any sound policy which Ambassador Lodge might develop." Touching a little closer to my father, though still without mentioning his name, it warned that his Montagnard troops could "desert the national cause" and upset the stability of the Highlands. It ended with a dig about the Bay of Pigs: perhaps the whole fiasco was "a desperate effort of those who helped lose Cuba for the Free World to try to recoup their loss of face."

The story hit Saigon early on Monday, September 2, 1963. It was a busy day for my father. In a series of cables to Washington, he reported on the aftermath of the coup. The generals were convinced Nhu was aware of the plot all along and they resented the Americans for limiting their support to encouraging words. Several sources were talking about Nhu's peace overtures to the North—one said that he'd even hinted that the "Northern brothers" would give South Vietnam a breathing spell if America cut aid. But other sources said Nhu was ready to compromise, even to the point of sending his wife out of the country. There was also a big meeting between Nhu and fifteen of the generals when he accused the CIA of wanting him "out of the way" and warned that the Montagnards and border scouts could be a weapon for Diem's enemies, another dig against my father. But Lodge was now on Diem's side, he insisted.

My father also sent a damage assessment that day. Although Diem and Nhu hadn't won a decisive victory so far, he said, we had to assume they were fully aware that the United States and the CIA encouraged the coup. The *Times of Vietnam* story was just a hint of how much intelligence they had gathered. "We estimate that they know good portion, possibly most of substantial details of our approaches." One positive result is that Diem could see how close he had come to disaster. Maybe things would finally start getting better.

But that night, President Kennedy appeared on television. Diem had gotten out of touch with his people and needed to take steps to revive popular support, he said. This would include "changes in policy and personnel."

In other words, get rid of Nhu.

That Friday, my father met with Nhu for two hours. Although his report doesn't give any hint of what he said to explain away the coup, the meeting seems to have been conducted with elaborate Asian civility. Don't worry about that silly assassination list, Nhu assured him. His brother-in-law had no agents and no power and wasn't going to kill anyone. He insisted he had nothing to do with the *Times of Vietnam* story or the attacks on the pagodas—Diem had approved martial law and the removal of the monks, asking the generals not to harm them. And all that stuff about negotiating with the North was lies too. He'd never go behind the back of the United States of America.

Then he changed the subject to his children. They were so isolated at the palace, he said, with too many servants and no playmates. Surely the wounds of this odd childhood would have lasting psychological effects on their adult lives.

Perhaps this was an effort to connect, to show his humanity.

And maybe it worked—which would explain why my father thought it was worth mentioning.

But he ended his report on a note of Aurelian detachment:

> See no point in trying to elaborate on his sincerity or insincerity but do not exclude that there are various substantial elements of deception involved in his statements. This would not be unnatural in power or politics.

That same day, he wrote his last letter to us. "As you have probably read in the newspapers," he began, "there has been a certain amount of excitement in Saigon but nothing extraordinary."

> At the moment, things have quieted down considerably and the city looks as charming and quiet as it ever did. Everyone here has taken recent developments in stride and morale is high. So far, the war against the Vietcong does not seem to be have been seriously affected, but various civil counterinsurgency programs have slowed down to a certain extent . . .

At the embassy, he continued, things were going on as always. Social life had slowed down because of the "slight tension and uncertainty" but tomorrow, he was hoping to get in eighteen holes of golf—his first game in the last three weeks. And he found it most pleasurable to work with Ambassador Lodge, "who continues to impress me as a very fine personality, highly intelligent, and with a genuinely patriotic desire to serve our country."

At the end, he slipped in a personal aside to my mother: "I found your letter from the New York Statler most heartwarming, but there is no reason to feel at all worried about any of us here."

In fact, he was under terrible pressure. Lodge had taken such firm control of the embassy that when Bill Colby arrived in Saigon that week for a fact-finding tour ordered by the White House, Lodge refused to let him meet with Diem or Nhu, saying it would send mixed messages about who was in charge. Clearly, Lodge still had set his mind on some kind of coup. In his cables home, he called Diem a medieval Oriental despot who could not even comprehend the idea of public service, who understood "few, if any, of the arts of popular government." Even when urged by Kennedy himself, he refused to meet Diem.

On September 10, my father made another fateful move. Sitting down alone at his desk, he composed a long cable reassessing the situation. Clearly Diem had underestimated the seriousness of the Buddhist crisis and let things drift out of control, he admitted. But he was right about one very pertinent thing. It now seemed clear that the core leaders of the Buddhist protests had "far-reaching political objectives" from the outset. Furthermore, Nhu's responsibility for the pagoda attacks was now in question too. Consider Diem's declaration of martial law, consider his request to General Don that the troops try not to hurt the Buddhist monks when removing them from their pagodas. Another Vietnamese general named Do Cao Tri had been making aggressive moves against Buddhists in the north for months. And certainly the failure of the coup had exploded the "often-held assumption that certain general officers and other dissidents would move quickly if given green light." Given all of this, my father said, it seemed likely that the generals would find a way to work it out with Diem. "Our impression is that there are few points of no return in Asia and that there seems to be a large amount of stretch in Asian societies."

Then it came time for my father to make the Agency's regular monthly payment to Tung's Special Forces. Although my father first "sought and obtained approval" from Roger Hilsman and the White House, as Hilsman took special care to point out in his memoir, an angry member of the Saigon CIA station leaked word of the payment to the press. The result was headlines all over the country and a much-syndicated Herblock cartoon that showed a bureaucrat giving a lecture in front of a map of Vietnam: "Basically, there are three governments involved—the Diem Government, the U.S.A., and the C.I.A."

The *New York Times* pointed out that Tung's troops were almost never used in combat but served primarily as Nhu's "private security force."

The *Washington Post* ran a fiery editorial slamming the CIA for taking on more than it could handle. "One would have supposed that the Bay of Pigs debacle would have alerted the White House to the risks of allowing an intelligence agency to sit in judgment on its own operational missions."

Two days later, a senator from Alaska made the first call for congressional oversight of the CIA—a battle that would continue for another thirteen years.

But in Saigon there was a new crisis. Now the protests were spreading beyond the Buddhists—first to college students, then to high-school students and beyond. One morning, Diem's policemen arrested more than a thousand protesting grade schoolers. Even the most prominent Vietnamese had to go to the police stations to bail out their children.

In his memoir, *Mission in Torment*, embassy press officer John Mecklin remembered the tension building into a "funny farm atmosphere." One day, he got so paranoid he carried a gun to work.

That's when my mother and sister and I got back to Saigon, just another trip through the airport for us kids, a chance to see our dogs and friends. But for my mother it was a plunge into a world gone grim. With friends like Marga Meryn, she grumbled about poor Fritz Nolting getting fired on vacation without even a chance to defend himself, how Lodge didn't come in till lunchtime and then went home for a nap, how his two adjutants shielded him from anyone who might voice a dissenting opinion. Her crowd had particular scorn for "those bloody reporters who think they know everything" and the young White House staffers who kept flying out to race around the country and prepare instant analyses of impossible problems. My father wouldn't listen to that kind of talk, she said.

But sometimes, when the pressure got bad and it was just the two of them, he'd grumble about all the young hotshots coming out to tell everyone what to do.

By then even Dave Smith—his deputy, who lived across the street with his wife and eight children—had turned against Diem and said so openly. "John was pretty sad about the whole thing," my mother remembered.

And it would soon get uglier.

28

•

REPLACE RICHARDSON

On the afternoon of September 11, Lodge sent a long cable to the White House. The situation was getting worse he said. With the arrests of grade-schoolers, the regime was now brutalizing children. The popularity of the United States was at stake. It was time for a "drastic change."

But at the White House, they still weren't sure what to do. At the Vietnam staff meeting that night, General Taylor asked how Abraham Lincoln would have reacted if a hostile religious movement rose up against him in the middle of our Civil War. The Secretary of State said we'd dealt with dictators before and shouldn't let that spook us into doing something rash. The Secretary of Defense urged caution too, suggesting that Lodge get out of Saigon and get to know the country better. McCone suggested a special envoy to talk to Nhu, since my father was becoming too controversial. And Kennedy tried to think of ways to get Lodge to open a dialogue with Diem—maybe if he wrote Lodge a personal letter, would that persuade him?

When the meeting ended, the Secretary of State telegraphed Lodge with new instructions—try "persistent talks" with Diem and somehow make him see that everything he had worked for was now threatened by failure. Again Lodge refused. "Do not see advantage of frequent conversations with Diem if I have nothing new to bring up," he cabled back. "Visiting Diem is an extremely time-consuming procedure, and it seems to me there are many better ways in which I can use my waking hours."

The next day, Lodge wrote a secret message to the Secretary of State. "I ask that you show this letter to the President personally, as it is vital that it not get into the government paper mill. For maximum security I am typing it myself and am sending it to you by messenger."

He wanted my father fired and replaced by Ed Lansdale, the legendary CIA man who helped establish Magsaysay in the Philippines. "This is said without casting any reflection on Mr. Richardson. Indeed I think of him as a devoted, intelligent and patriotic American. If his loyal support in the past of the U.S. policy of winning the war with Diem has made it difficult for him to carry out a different policy now, he has never said so or showed it."

Later, some of the men involved would puzzle over why Lodge did this. Hilsman said it was less personal animosity or rivalry than a diplomatic signal to Diem and Nhu. Halberstam said that my father's presence stopped intelligence from "moving freely" and put the generals in doubt about the change in American policy. Others pointed to the influence of Rufus Phillips, the former CIA and AID officer who had become one of Lodge's unofficial advisors in private meetings, he kept praising Lansdale to Lodge in private meetings. In public, none of them wanted to touch on Lodge's prickly imperiousness or my father's deviations from perfect obedience—the "second thoughts" incident and the cable when he reconsidered the wisdom of the coup.

In any case, McCone vetoed the idea.

Later, Lodge vented his feelings in a revealing letter to the Secretary of State: "You can be sure I will continue to do my very best to carry out instructions even if I just use persons trained in the old way, who are widely (and however unjustly) believed to be in touch with those who we are trying to replace and who, without ever meaning to be disloyal, do in fact neither understand nor approve of current United States policy."

A tiny epic of resentment couched in tones of upright diplomatic reason, this letter lights up my father's dilemma like a strobe.

Over the next few days, the power struggle kept bubbling away. Acting on a tip from a "senior CIA official" who had just come back from Saigon, McCone told the President that Lodge was thinking about encouraging General Don to mount another coup attempt.

Shortly afterward, the Secretary of State told Lodge not to stimulate any coup plots "pending final decisions which are being formulated here."

Two days later, Halberstam published a piece about tension inside the embassy. Then a UPI columnist named Lyle C. Wilson called the United States mission to Saigon a "five-headed monstrosity." The *Times of Vietnam* then ran an exposé of the "pro-coup and no-coup factions" at the Saigon CIA station under the headline PARDON, CIA, YOUR SPLIT IS SHOWING. Through all this my father kept working, sending home reports that were striking, especially given the ongoing charges that he was too close to Nhu, for their objectivity. In one, he gave another detailed rundown of Nhu's peace negotiations with North Vietnam, a subject particularly alarming to Kennedy and his advisors. He also blamed Nhu for encouraging one of his lieutenants to intimidate Lou Conein with a gun. Despite his demurrals at their last meeting, Nhu still seemed to have hard feelings about the CIA's involvement in the coup plots.

Another cable reported on a meeting with General Khiem, who was starting to look on the bright side. He thought that Diem had finally turned the corner, that he was accepting his mistakes and making efforts to control people like Colonel Tung. Once he gave the generals more power, surely Nhu's influence would wane to a minimum.

Instead of cheering, my father was skeptical: "We do not share Khiem's belief that Diem will accept and introduce recommendations made by the General Officers to the extent hoped for by Khiem."

But these reports were secret, and the press was herding in the opposite direction. On September 17, the latest attack appeared in the *Washington Post* under the headline U.S. POLICY MIRED IN VIEWS OF 3 AGENCIES, yet another story that faulted the CIA for ignoring Diem's mass arrests and unpopularity. By noon, the Secretary of State was asking McCone if it was time to bring Richardson back for "consultations."

Okay, McCone said. If Lodge wanted a new station chief that badly, they could replace Richardson—but not with someone from the "outside."

That killed the draft-Lansdale movement, since Lansdale no longer worked for the agency.

By this time, the CIA was starting to get suspicious about all the bad press my father had been getting. In the blunt words of an internal study the Agency conducted later, "Ambassador Lodge had begun to criticize CIA Station Chief John Richardson sharply, and word of this development soon appeared in the press." John Mecklin painted a similar picture in his memoir, telling how Lodge arrived in Saigon with instructions that no-body in the embassy would talk to the media but him. "The leak is the pre-rogative of the ambassador. It is one of my weapons for doing this job." Mecklin's job as embassy press officer put him in an unusually good posi-tion to diagnose the flood of five-headed-monstrosity stories, which all revolved around the idea "that strong leadership was needed and Lodge was the man to provide it."

There was another candidate. On September 26, John McCone had lunch at his home with James Reston of the *New York Times*. According to a memo he wrote immediately afterward, Reston told him the campaign of leaks had been "obviously planted . . . probably a good deal of them from Harriman." Given Harriman's strong will and his role as the prime mover of the Cable of August 24, this makes sense. And McCone thought the point worth elaborating in another memo later that day. Because the agen-cy's emphasis on caution had become exasperating to "those who wished to move precipitously," he wrote, said persons were now retaliating by "carry-ing on a campaign against the CIA and the Station."

Probably it was both of them, and others too, the permanent whisper-ing campaign of Washington doing its stealthy work. The irony is that my father seems to have been moving steadily to a negative view of Diem and Nhu. Just one day after McCone's lunch with Reston, for example, the Secretary of Defense spent two hours interviewing him in Saigon and then put down his "words and sequence of thought" in a detailed memo to Kennedy. Although most of the meeting seems to have been given over to a frank report on the very gloomy views of the Vietnamese Secretary of State, who blamed Nhu for the pagoda raids and said Diem was dragging his country into disaster, the first five points seem to reflect my father's personal views:

1. The Buddhist crisis crystallized the discontent which had been dormant for some time.
2. The future is so uncertain he cannot predict what will happen.
3. The massive arrests of the students have been very bad.
4. The arrests include the arrest of children of the military officers and high-ranking bureaucrats.
5. The night arrests are particularly bad for they cause people to hate you.

But it was too late. Around the same time, Lodge asked my father to consider suspending payments to all CIA projects that served "politically repressive activities" and to consider acquiring a big stash of Vietnamese money in case they wanted to bypass Diem's officials and make direct payments to the counterinsurgency programs. In response, my father noted that the number of CIA programs Diem was using for repressive political activities was "actually rather limited" and cutbacks would have a pin-prick effect. As to the idea of making direct payments to the counterinsurgency programs, he didn't see how to do that unless a local Vietnamese commander decided to defy the central government.

Four days later, Lodge wrote some bitter notes in his private papers. "Station should be out of Chancery building . . . head of Station *not* to be Special Assistant to the Ambassador . . . should end backgrounding of press by CIA people of high or low degree . . ."

Still not satisfied, he expanded the theme: "CIA has more money; bigger houses than diplomats; bigger salaries; more weapons; more modern equipment. A Joint Congressional Committee on CIA?"

The next morning, my father's life exploded.

29

•

CIA CHIEF RECALLED

The bombshell appeared in the *Washington Daily News*, written by a reporter named Richard Starnes: THE CIA MESS IN SOUTH VIETNAM— ARROGANT CIA DISOBEYS ORDERS. A little-known writer who sidelined in detective novels, Starnes came out to Saigon for a few weeks and quickly stumbled onto a very inside story indeed—that the head of the CIA in Vietnam was a man named John H. Richardson and that he had twice refused to carry out explicit orders from Henry Cabot Lodge. Denouncing the CIA's performance in Vietnam as "a dismal chronicle of bureaucratic arrogance, obstinate disregard of orders, and unrestrained thirst for power," warning of "spooks" and "button men" who were accountable to no one, Starnes packed his article with a surprising number of specifics. In one case, the CIA "frustrated a plan of action Mr. Lodge brought with him from Washington, because the Agency disagreed with it." This led to a dramatic confrontation between Lodge and my father that even McCone and the Secretary of State couldn't resolve, putting it on President Kennedy's desk. Starnes also quoted military men who were bitter about the way "spooks" dabbled in military operations and State Department aides who thought the CIA was supposed to gather intelligence, not make policy. He quoted anonymous CIA agents who swore they warned that Nhu was going to raid the pagodas, but the information "wasn't passed to top officials here or in Washington." That was why Washington stumbled afterwards. And don't forget that the CIA was *paying* Nhu's Special Forces, and when top officials here and at home expressed their outrage, the CIA *continued the payments*. "It may not be a direct subsidy for a religious war against the country's Buddhist majority, but it comes close."

And with his novelist's flair for drama, Starnes pushed his story right to the archetype that lurks behind all CIA legends, a *Seven Days in May* scenario where a shadowy secret organization subverts the entire democratic system. In the words of yet another of his anonymous sources: "One high official here, a man who has spent much of his life in the service of democracy, likened the CIA's growth to a malignancy, and added he was not sure even the White House could control it."

When the article hit Saigon, my father was stunned and dismayed. Exposed by name in the press, accused of insubordination, it was about the worst thing that could have happened. His co-workers were just as shocked as he was, and later Mecklin remembered the buzz in the embassy hallways about whether Richardson had been insubordinate, the common feeling that it couldn't possibly be true. "I was a witness on one occasion when Lodge gave Richardson an order that I knew he strongly disapproved. His response was, 'Yes, sir.'"

But his reports continued without a ripple, in the same tone of measured detachment. In a cable about a meeting between Lou Conein and General Don, he reported that the generals finally had a coup plan. He got Lodge's approval for a follow-up meeting with Big Minh and reported the plan: Conein would just show up at Minh's office in an army uniform, pretending to be there on an official visit. If anybody asked, he was there to discuss where to put the new Green Beret headquarters.

But the Starnes story couldn't be contained. A day later, Max Frankel followed up on the front page of the *Times*. "President Kennedy was reliably reported today to have recalled 'for consultations' the head of the Central Intelligence Agency operations in South Vietnam, presumably to end his policy dispute with Ambassador Henry Cabot Lodge."

In the second paragraph, Frankel used my father's name.

In an accompanying column, Halberstam contributed a savvy local analysis. Not only was there "no evidence that the C.I.A. chief has directly countermanded any orders by the Ambassador," not only was there a consensus that my father had done a superior job in a difficult situation, but some warned of "a growing effort to make the C.I.A. the scapegoat for the unhappy events of the last six weeks." What's more, Lodge's fight against him seemed to echo his fight with General Harkins over who was going to run the show in Saigon.

But my father's name appeared in this story too.

Soon it was everywhere, shooting from newspaper to newspaper all across the world. Following Frankel, many of the stories suggested that Lodge wanted a replacement who would confine himself to gathering intelligence, which is hilarious when you consider how eagerly Lodge used CIA assets to push his coup. But it was the first time a CIA station chief had been publicly identified, and the press was giddy with it.

On the afternoon of October 5, a monk named Thich Quang Huong sat down outside Saigon's central market and lit himself on fire. As the Saigon police moved in their barricades and the army rolled in its tanks, my father filed what I believe is his last cable from Saigon:

> Lt. Col. Conein met with Gen. Duong Van Minh at Gen. Minh's headquarters on Le Van Duyet for one hour and ten minutes morning of 5 Oct. 1963. This meeting was at the initiative of Gen. Minh and had been specifically cleared in advance by Ambassador Lodge.

The generals were now thinking about assassinating Nhu and leaving Diem in office, he said. Or maybe they'd surround Saigon with troops and mount a siege. Or fight their way into the city. Minh asked Conein what the United States would say and Conein said he would have to get instructions.

With that, my father got up from his desk and headed down the hall. He was planning to leave without saying goodbye to Lodge, a rare breach of protocol and another glimpse of his hidden anger. But one of Lodge's aides caught him and said the ambassador wanted to see him. So he headed back to Lodge's office and they had a brief, stiff conversation. "You're tough," Lodge told him. "You can take it."

At home, he told my mother only that he'd been called to Washington. He reminded her to check the safe for security instructions in an emergency. Then his driver took him away, alone.

We stayed on, waiting for him to come back. Later my mother would be bitter because Dave Smith—his deputy, now his replacement—didn't tell her what was happening even though he lived right across the street, even though my father had always been kind to him and never stopped him from expressing his views. Instead she heard the story from a couple of generals who came over to say how sorry they were. She kept up on the news in the clips and teletypes that John Mecklin sent over from the embassy, headlines like CIA CHIEF RECALLED running side by side with pictures of Thich Quang Huong burning himself to death.

When my father reached Washington, the CIA stashed him at a friend's house to keep him away from the media. A few days later at a meeting of the Senate Committee on Foreign Relations, a conservative Republican senator named Bourke B. Hickenlooper asked the Secretary of Defense if there had in fact been any dispute between Lodge and Richardson. When he dodged the question, Hickenlooper demanded a "point-by-point refutation" of the shocking and false charges in the article by Richard Starnes. But the stories in the newspapers kept getting more and more specific. On October 8, the *New York Times* published a feature by Malcolm Browne that reviewed the whole story and offered history's first glimpse of a real CIA station chief: "Mr. Richardson is bald, wears glasses with heavy horn rims and dresses smartly and conservatively. He looks and acts every inch the diplomat. He is a specialist in counterinsurgency, having fought Communist guerillas for his agency in Greece and the Philippines."

But my mother could barely get past the lead, which recapped the story she told Browne at the New Year's party about hiring Buddhist monks to exorcise our ghosts, dressed up with a few ironic details about the French secret police using the house to torture political prisoners.

Then Mecklin began sending over the reaction stories, including a transcript of an NBC radio piece by Morgan Beatty calling my father a "guerrilla servant extraordinaire" and "a remarkable diplomat with an engaging personality." Clearly someone inside the agency had been feeding Beatty, because the piece included a vivid description of the Montagnard program and also the pertinent (if not yet completely true) news that it had already been turned over to the military. Mecklin also sent over an editorial from the *Washington Evening Star:*

> The crime Mr. Richardson is said to have committed is truly fascinating. He is being accused in the bars of Saigon with declining to overthrow the government of South Vietnam. This is a charge so strange we can hardly credit it.

But the negative stories kept coming, including a particularly indignant *New York Times* editorial that accused the CIA of acting as "a state within a state" and my father of acting like a "kingmaker." In the *San Francisco Chronicle,* he was "attempting to run American policy." In the *Washington Post,* Drew Pearson practically accused him of treason: "The real reason for CIA Chief John Richardson's recall from South Vietnam was the discovery that he had been reporting to President Diem's ruthless brother, Ngo Dinh Nhu, what opponents were saying about the government."

In these stories, Lodge came off like a genius. In the *Chronicle,* he "never appeared more shrewd, skillful and operationally sound than he did when he asked for the recall of the CIA chief." In the *Post,* he sussed out Richardson's scurrilous leaks to Nhu and immediately arranged for "direct orders from Washington" to stop him. Clearly, somebody was feeding the other side too.

Then Senator Eugene McCarthy started pushing a bill to set up a congressional committee to watchdog the CIA.

And finally someone put the question directly to Kennedy: Had the CIA been acting "independently" in Vietnam? This was at Kennedy's regular televised press conference, which happened to fall on the day before my father's 50th birthday. Kennedy said he'd looked through the record carefully and the CIA had done nothing but support policy.

When news of this speech reached Saigon there was "widespread satisfaction" in the embassy, Mecklin said in his memoir.

But for my father, it was the sound of a career ending.

My mother kept busy wrapping things up. She found a place for Mercy and said goodbye to all her friends. She gave me a birthday party, making it special by hiring an elephant from the local zoo. And just before we left, she got an invitation to a party at Lodge's house and refused to go, a shocking bit of defiance for someone who had worn so many pairs of white gloves. Let "Henry Cabbage Lodge" spin for all she cared.

Then she got a call from Lodge's office. The Ambassador really wanted her to come because she knew a lot of the people, his secretary said. She could be helpful. So she went and filled a plate at the buffet and sat down on the screened porch with a *Time* magazine correspondent, and Lodge sat down with them and tried to make a joke. "Everybody says that Cabots talk only to the Lodges and the Lodges talk only to God, and here I am talking to all you nice people." Then he smiled and put his hand on the reporter's knee.

My mother never forgave him. "Bastard," she said. "He was the most arrogant man in the world."

Four days after we left, he moved into our house.

30

●

VIRGINIA, 1963

We arrived at Dulles at two in the morning, bleary eyed from the long trip. Nobody was there to meet us. My father was still in hiding and my mother didn't know how to reach him. There were no taxis and just one ticket counter open, and at first they didn't want to let her use the phone.

She called Bill Colby at home, waking him up. "You're not supposed to be here till tomorrow," he said.

It was a mix-up, something to do with international time versus Washington time. "How can I reach John?" she asked.

"I can't tell you. But I can call him and give him your number."

So she gave him the number of the ticket counter and hung up, and a few minutes later my father called to say he was sorry nobody had met us. She should take a taxi to a hotel. He'd contact us again in the morning.

While we waited for the taxi I sat on the stairs, excited by the long trip and unusual situation, talking to everyone who came down the ramp. What are you doing here so late? Do you work for the government too? Finally one man asked me what I was doing there in the middle of the night.

"Waiting for my daddy, John Richardson."

"Oh," the man said, moving on quickly.

The next few weeks were chaotic. Marga and Sam Meryn found an apartment for us, two bedrooms filled with rented furniture. School had already started, so they scrambled me and my sister into the Congressional School in Falls Church, where we wore uniforms and realized that we were not like regular American kids. We knew how to bow and curtsy. We talked like little adults. We didn't know how to cuss and knew nothing at all about pop culture. But there was so much to absorb we barely noticed. Just living in a building with elevators and halls was strange and odd and so full of new stimuli—television, radio, lawns, parkways. And everybody spoke English.

One day my mother drove us across the Bay Bridge—my father riding next to her with white knuckles, frozen by acrophobia—and we visited Frank and Polly Wisner at their place on the Maryland shore. My mother was almost forty now but still felt too young (or too something) to fit in with the senior CIA wives. But they were all there and so were their husbands, come to show their support. After one or two speeches, my father started to get up and Frank put his hand on his shoulder. "You don't have to reply, Jocko."

But my father got up. "I learned a lesson in Greece," he said. "Greeks like to get up and talk."

And he thanked them all, milking the comedy in his tendency to ornate oratory, striking fresh notes of philosophy and resignation.

They gave him a new job, Deputy Director of Training. Every morning he left for Blue U, the blue-tiled building in Arlington where old spies trained young spies. The understanding was that the boss was scheduled to retire soon and then he would become Director of Training, which would put him in charge of one of the CIA's four divisions.

He didn't care. The only job he ever wanted was station chief. And his mind was still in Saigon, where his absence was making the generals bold. "We were fortunate when Richardson was recalled," General Don wrote in his autobiography. "Had he been there, he could have put our plan in great jeopardy." Now no one wanted to slow things down and no one gave a contrary opinion. When Kennedy zigzagged, as he did repeatedly, Lodge kept pushing. When Diem made a desperate last-minute offer to capitulate completely and do everything the United States wanted, Lodge buried the news in a low-priority cable. And when troops attacked his palace and Diem made one last phone call to the embassy, asking in his stiff and formal way for a statement on the position of the American government, Lodge just dodged the question. "But you must have some general ideas," Diem pleaded. "After all, I am a chief of state. I have tried to do my duty."

And Lodge dodged the question again, telling Diem how much he admired his courage. "Now I am worried about your physical safety," he said.

On the morning of November 2, as rebel troops approached the palace, Diem and Nhu slipped out and went to church. One of the generals later remembered waiting in the office of General Le Van Ty with Lou Conein and a briefcase holding $150,000 in piasters, payoffs to Don and others for the coup. When news came that Diem and Nhu had committed suicide, Conein went pale and stood up, cursing and telling them they'd all be held responsible.

Rufus Phillips told me that Conein was still pale and upset when he saw him a few hours later. "The goddamn bastards," Conein said. "They murdered them."

Later the real story came out. Snatching Diem and Nhu right out of church, the rebel soldiers pushed them into a truck and tied their hands behind their backs. At a railroad crossing, they stopped the truck and shot them in the back of their heads. Then they shot them some more and stabbed them too.

This was almost certainly done on the orders of Big Minh, our new ally.

Because of CIA rules about closing the door on past jobs, my father probably wasn't able to get many of these details. But that night, he had dinner with Fritz and Lindsey Nolting and Bill and Barbara Colby for what Colby later described as "probably the only Washington wake for the Diem regime and the Ngo brothers." This is his account of that night:

> We all joined in expressing concern over how Vietnam—
> and the United States—would face the problems ahead with-
> out Diem's strength and leadership, and in wondering how
> our Government could have been so blind as to have con-
> tributed directly to his overthrow and death.

Less than two weeks later, ten thousand Vietnamese students took to the streets of Hué to protest again, this time against the military junta. And soon Buddhists started burning themselves again. And then a small article appeared in the paper saying that General Tung had been executed in Saigon. And not long after that, a few days at the most, my father got a letter with a Saigon postmark. It was from Tung, written just before his death. He said he was sorry that he and his colleagues had gotten "an unfair negative impression" of him, sorry that they believed the article in the *Times of Vietnam* saying he had conspired in the coup. True friends were rare and should be honored.

My father threw the letter away.

During all this, I barely noticed my parents. I drifted through the world in a kind of fog, observing this and noting that and letting it all slip away without ever really knowing what any of it meant. At that moment the only thing real and important was:

1. Silly Putty. It comes in that cool little flesh-colored egg and when you squeeze it, it pops and crackles and you can push it down on newspaper cartoons and pull up the drawings like magic, color and all. I dream of Silly Putty. I *crave* Silly Putty. I want Silly Putty so bad I have begun to live a secret Silly Putty life, riding my bike past the little stationery store just to be in its presence. And one day when I'm riding there I hear a song on a car radio that makes me stand on the pedals—*Oh yeah I, I tell you something, I think you'll understand.* It's so fresh and cheery and driving, like nothing I've ever heard before, and somehow it mixes up with how much I want that Silly Putty and suddenly I feel so alive it's like something bursting inside my chest.

But life went on, strangely bland. At cocktail hour my sister and I sat down with our parents and gave our reports on the day, then went back to our rooms to do homework or into the living room to watch TV. My favorites were *The Three Stooges* and *Hogan's Heroes*. One day my father came in to the living room angry because he objected to *Hogan's Heroes*. It was a mockery, he said, an outrage. There was nothing funny about a Nazi prisoner-of-war camp and anyone who thought there was didn't understand a goddamn thing. But that's my only memory of him in those months. He worked late and kept to himself. But I do know that one time he went fishing with Jim Angleton, who brought along a pair of secret spy glasses that helped him see the trout. Another day, the Agency put on a small secret ceremony to give him a Distinguished Intelligence Medal for his service in Vietnam. And one time he took us to visit a place my parents called the Farm, a nice woodsy place down below Williamsburg where my father's team taught young spies how to plant bugs and interrogate suspects. For some reason the trip made a big impression on me.

22. I'm standing in an old general store with wooden bins where they sell these crackly onionskin copies of the Constitution, not the real thing but convincing copies that I *must buy* for the kids in my class. As my mother looks on in surprise, I buy four of the stiff sheets and wait like a grownup while the owner puts them into a tube.

23. In her bedroom, my mother tells me she's missing twenty dollars from her wallet and asks me if I took it. When I tell her about wanting to buy gifts she seems puzzled for a moment and then gives me a curious look—and shame floods me. She has seen right through me. I was trying to buy friends.

That might have been around the time my father invited Frank Wisner to give a speech to his trainees. The way my father would remember it later, they didn't respond at all. They were just bored. Here was the man who had founded the political action arm of the Agency and started Radio Free Europe, who had driven himself to the point of a breakdown twice, and they didn't care. They were a new generation.

Frank was hurt. My father was sad.

In the spring, he paid four hundred dollars to join the Riverbend Country Club, a nine-hole golf club started with seed money from other spies living in the area. My sister and I joined the swim team. We spent many days freezing in our Speedos at dawn, thrashing at the water and winning ribbons while my father played golf with George Mistowt and Sam

Meryn, who had come to work for him at Blue U. And Mason and the Messings were living nearby now and the Messings sang in choral groups and Mason's son—later he'd go to prison on a drug charge—was just starting his life as a juvenile delinquent. I remember a blur of houses in the country and different kids, Risk and Monopoly and muddy riverbanks, crepes suzettes in the morning and fields of high grass that had to be cut with a tractor.

> 24. And one boy's sister really wants to play Monopoly with
> us and promises to come after the parents go to sleep.
> And she does and we play Monopoly for a long time
> with a weird lazy intensity and then she lies down and
> covers herself with a blanket, waiting for me to join her.

One day, Lou Conein asked my mother out to lunch. This was odd. Elyette Conein had helped her many times, from finding the house in Saigon to helping her decorate Bill Truehart's living room so he could have official parties, but she'd never been particularly friendly with Lou. At a restaurant near McLean, he told her how isolated my father was in Saigon toward the end, with even Dave Smith against him. But his main point seemed to be about Lodge. In the wake of the murder of Diem and Nhu, Kennedy and Lodge both made statements about how shocked and dismayed they were, Lodge insisting he had no idea Diem would be shot. But that wasn't true, Conein said. Lodge knew they were going to get killed. He promised to send a helicopter to help them but never intended to do it. She wasn't exactly sure why Lou was telling her these things. "Maybe he was getting a conscience," she told me later. But you couldn't always believe the things Lou said.

> 25. And now we're in West Virginia on a farm with a pond
> and snapping turtle and a girl my age. While the
> grownups are up at the house talking, we hide behind a
> hedge and bend toward each other to see what a kiss is
> like. At the end of the day we say goodbye with a desper-
> ate forsaken hug that makes the grownups raise their
> eyebrows.

Then my father found a lot right next door to CIA headquarters and dreamed up a big white house on a hill, which came as a big surprise to my mother since he always said from the very beginning that he didn't want to own a house, that he gave up the settled life when he joined the Agency. But something had changed his mind. He gave the CIA a commitment to spend four years as director of training and told us he wanted us to "put down roots in America." So they had an architect make the drawings and we moved to a big house near McLean to wait out the building process.

That was my first suburban summer. I remember finding box turtles in the woods and a little pond where I would toss out a line and hook sunfish. I spent a lot of time camping with the Boy Scouts, which my father thought would help develop my love of nature. And one day my mother stuck a white slab of something called "suet" near the window and got all excited about a bird she called a "pixilated woodpecker," words I marked in my memory as if they were yet another strange foreign phrase I had to fix or lose forever.

But we had trouble adjusting. My mother was alone for the first time, stuck in the middle of nowhere with two small kids and no servants. One day she started a fire in a pan of bacon and tried to put it out with water and ended up burning down the kitchen, and afterwards my father took her to an agency to get a live-in maid. Then she got a job selling advertising for the first issue of a new magazine called *The Washingtonian*. She quit to pick out doorknobs and cabinets for our new house. Then my sister and I had to switch schools again, so she drove me out to a place called Landon, a woodsy prep school for the children of generals and senators. For Jennifer they chose the Madeira School, where students who wanted to bring their own horses to school could keep them in the campus stables. But Jennifer said she didn't want to go to another school where she had to wear pleated skirts and she definitely didn't want to go to a girls' school. She just wanted to go to a regular public school where she could be normal. Although I liked the look of Landon's vast green fields and ivy-covered buildings, I joined her rebellion out of general obstinacy.

At McLean Junior High, Jennifer started out shy but thrived in dance class. I was a bit less successful at making the transition, as they say. Introducing myself in class on the first day, I said I had "been abroad" most of my life, which was the way my parents always described it. Everyone laughed. On the playground later, I got punched in the face. "You were awful," my mother remembered. "You were practically flunking everything, plus you were acting like a little smart-ass. The teacher called me and said you thought you were better than anybody else. The other kids hated you."

In January, I transferred to Landon.

But my father was still pining for an overseas assignment. I learned this many years later, sitting at the CIA computer at the National Archives. Among the millions of declassified documents are diary entries describing conversations with Matt Baird, his supervisor. "Matt says that deep down, Jocko really wants to go back overseas," says one dated January 25, 1965.

A week later, there's another:

"On Matt's advice, Jocko has talked with Dick Helms about a possible overseas assignment. Dick held out some hope but did not wish Matt to postpone his retirement."

On a crisp Sunday morning that fall, my father went out to the back-yard to split wood. For a few minutes the sound of vigorous chopping rang through the woods—then the sound stopped and he came in through the back door and flopped down on the sofa, grabbing his shoulder. He felt a terrible pain, he said. Then he fell off the sofa.

My mother ran to the telephone and called Dr. Mistowt, who told her to get him in a car and bring him to the hospital right away. "I'll meet you at the emergency room door." But when she hung up and turned around, my father was gone.

She found him in the bedroom, holding a sock. "Come on, come on," she said, "I've got to get you to the hospital."

"Just a moment, I can't find the sock that matches this."

At that moment Sam and Marga Meryn arrived, bringing Doris to spend a day with Jennifer. My father managed to climb in the backseat of their car and lay there shaking as my mother put a blanket over him. When they got to the emergency entrance, Mistowt was waiting with a gurney and a cardiologist.

It was just as he predicted, just like his father: a heart attack. For the first few days, someone had to shave him and feed him his meals. We went in once a day for a half hour and I have a blurry memory of seeing him weak and helpless on the bed and feeling vaguely responsible for it. In the memory I see myself sitting there with long hair, mixing an injury I inflicted with one I didn't.

He spent the next month living in our basement apartment, too weak to climb the stairs. When he finally started to get better, he took me to plant flower bulbs in the yard of our new house. They had just painted the bricks white and the hill out front was still bare dirt. He wanted me to help him and be involved, but I was mopey about doing the work and he got annoyed. I'm afraid that none of it was real to me. The house and the heart attack were as vague as the beautiful flowers that were supposed to be inside those dirty bulbs, just another story told by adults.

Most of that summer, my father spent reading—his first chance to hit the books in two decades. Only now all the books were about Vietnam. That was the time of *Mission in Torment* and *The Making of a Quagmire* and *The New Face of War* and even Vietnam novels like *The Ambassador*, which was written by Morris West and set in Saigon as the Americans maneuvered to overthrow a hateful dictator. The ambassador is a noble fellow with an interest in Buddhism who participates in the coup only after getting assurances that nobody would be assassinated. The CIA chief is quite another story: "His voice was soft. His manner was full of charm and deprecation. He was reputed to be devious, ruthless and without moral scruples of any kind."

His name is Harry Yaffa, and he's an enthusiastic plotter with a "spoon in every pot" and a marvelous facility for keeping them all stirring at once. He loves intrigues and laughs as he discusses the possibilities of assassination. In one typical speech, he tries to explain life to the idealistic ambassador:

> "Let's be frank with each other, sir. You are the official representative of the United States. I have to serve in another way—as a political opportunist. There are things I have to do that you can never approve, and therefore it is better that you should not know them. I have to kill men and suborn women. I have to foment one plot to ensure the success of another. I have to provide against your success and your possible failure. If you want to salve your conscience by having me lie to you, I can do that, too. I'm very good at it, but I'd rather not lie when it isn't necessary. I hope I make myself clear."

"Very clear, Mr. Yaffa, except in one particular. What about your own conscience?"

"A luxury, sir. I found long ago I could not afford it."

Around that time, Frank Wisner took my father out to lunch. This was just four months before he killed himself. After a long talk about the books, Wisner admitted that he was afraid he was losing control and couldn't bear the thought of being institutionalized again. After the lunch he wrote a thank you note, one of the very few personal mementos my father kept. "You were in a good position to appreciate and sympathize with the frustration and restraint imposed by hospitalization, for your period of enforced rest was even longer than my own," he began. Then he turned back to the happier topic of the books, enclosing a copy of a review of *The Ambassador* that had just appeared in the *Catholic Standard*. Although the byline is Harold L. Scodly, the review was most likely written by Wisner himself. "As you are aware," he said, "this piece is part and parcel of my campaign to repair and improve the IMAGE of the apparat and was done with the knowledge and blessing of those at the top who were likewise concerned about *The Ambassador* and its potential for evil."

> As far as Yaffa is concerned, the author might as well call him John Richardson, in light of the fanfare of publicity which surrounded Richardson's enforced departure from Vietnam and his return to Washington. Only those readers with some discernment and retentive memories will recall that Richardson departed Vietnam well before the coup, and that his chief "crime" appears to have consisted of counseling caution concerning the viability and the aftermath of certain drastic remedies that were prescribed by others.

Always the student, my father read the Vietnam books with a black pencil, underlining passages in his careful hand. In *Mission in Torment*, he underlined jokes like "no Nhus is good news" and observations about the players, that Diem was "a geniune patriot" whatever his faults but also "a Messiah with a persecution complex." He put two checks beside the passage where Mecklin defended him. "Richardson was eminently successful in cultivating Nhu's confidence and respect. The subsequent press attacks upon Richardson, accusing him of being 'too close' to Nhu, were unfair and unwarranted. That was his job." In *The New Face of War*, he underlined the passage when Browne wrote about the battle that turned the course of the press war. "For the Viet Cong, Ap Bac became the victory cry."

Quagmire was the book he underlined the most, perhaps because Halberstam went after him so many times it almost seemed like a fixation. "Richardson enthusiastically believed in the policy" and the CIA was "actively on the team," and if some Vietnamese were wary of them, that's what happened to an agency "whose top members were part of the team." And of course this must have affected his reports, Halberstam speculates. "No man, once his career and prestige are committed to working for the success of a policy—and once he has told his superiors that such a policy will work—retains his objectivity . . . consciously or unconsciously, intelligence becomes tailored to fit the policy." And don't forget that Richardson "longed for control" over the American reporters, a charge Halberstam would later repeat in other forums although it seems to have been based on a single ironic remark. And worst of all, he was fatally wrapped up in Ngo Dinh Nhu. "In staying close to Nhu, Richardson had to pay the price of taking Nhu's word for an event, even of seeing the situation through his eyes. Consequently, it would have placed his position with Nhu in jeopardy if his agents were working the other side of the street and gleaning intelligence—as they should have been—from anti-Nhu sources."

At this point, my father couldn't restrain himself. In the margin, he wrote a single word: "untrue."

He skimmed past the pages praising Lodge, underlining only the line where Halberstam called him "the best ambassador we have ever had there." He paused longer over a passage that said Diem was acclaimed in his early days "not because of his character or his ability to govern" but because he crushed the gangsters and brought order to the country. This time he wrote in the margin: "Nine years in a country at war."

Toward the end, he underlined almost every line of the long section where Halberstam tried to look to the future. More than the personal criticism or the hagiography of Lodge, this seemed to get under his skin. "The basic alternatives for Vietnam are the same now as they were in 1961 . . . the first step toward a neutral Vietnam would undoubtedly be the withdrawal of all U.S. forces in the country and a cutback in military aid; this would create a vacuum so the Communists, the only truly organized force in the south, could subvert the country at their leisure."

In the margin, my father wrote: *Agree.*

What about withdrawal? "<u>It means that those Vietnamese who committed themselves fully to the United States will suffer most under a Communist government, while we lucky few with blue passports retire unharmed; it means a drab, lifeless and controlled society for people who deserve better. Withdrawal also means that the United States prestige will be lowered throughout the world, and it means that the pressure of Communism on the rest of Southeast Asia will intensify</u>."

Agree, he wrote again.

The only other possibility Halberstam could see was sending U.S. troops to fight on the ground, which would lead to "<u>a war without fronts, fought against an elusive enemy, and extremely difficult for the American people to understand</u>."

And my father wrote in the margin, this time underlining his own words to express amazement: "<u>He proposes no solution. He is against the only 3 alternatives he lists; neutralization, withdrawal and troop commitment</u>."

There was one last note. At the very end of the book, Halberstam ridiculed the "cliché" of Vietcong terror. "Can the guerrillas really create more terror than Government troops with howitzers, armed helicopters, armored personnel carriers and bazookas?"

In the margin, my father answered: "Yes they can."

31

DIRECTOR OF TRAINING

And finally, moving into that big white house on the hill at the end of Basil Road, we came into some kind of normal American life. Dad went back to working late and playing golf on weekends and Mom bought a spanking new Chevrolet Impala for three thousand dollars and drove my sister to ballet lessons and voice lessons while I rode in the back, soaking up hints of my sister's glamorous teenage world—songs like "Hang On, Sloopy" and "What's New Pussycat?" I met a kid named Bobby who played the drums and another kid named Chris and the three of us explored the woods near the CIA, gradually extending our range across the highway to the Potomac River. We caught crayfish and squished our feet in the mud at low tides, walked the mossy rocks to a little island. We built fires the high tides washed away. One day an old black man showed me his secret bait, balls of white bread soaked with a splash of vanilla extract.

When September rolled around, I put on a jacket and tie and went back to Landon. But soon there were fresh signs of trouble, and a tweedy old teacher named Irving Ricker began writing notes to my father fretting about my poor grades and complete lack of discipline. I was not working up to my potential, he said. I was more interested in social activities than schoolwork. Gusts of imagination kept sweeping me away from my goals. Then my grades started to plunge—a D in math, a D in Social Development, a D in General Effort, even a D in music.

As always, my sister and I gave our reports at cocktail time then went on about our business, usually going into our rooms or down to the basement to watch TV. At dinnertime, we'd gather at the Hepplewhite table for a formal dinner and talk about the issues of the day. My sister and I were always expected to "hold up our end of the conversation," as my father put it. It was part of our social training, like learning to bow and curtsy. As my bad grades piled up, however, these rituals began to strike me as more and more artificial. Trapped at the dinner table, I bristled at being forced to justify myself.

But the clearest memory of my father in those years is a happy one, a trip we took to the Blue Ridge Mountains. For some reason it was just the two of us, and I remember feeling surprised and privileged and a bit apprehensive. We stopped at a scenic overlook and ate hard-boiled eggs and he taught me the word "panorama," enunciating it as if it were a phrase in a foreign language, which fixed it so hard in my memory I can still see the trees and the rest stop and the salt in the tin foil that held the eggs.

After another semester, Ricker had become increasingly annoyed. In letters and notes to my father, he said I was impulsive and careless and took shortcuts and stumbled over words and required close supervision at all times and didn't seem to realize that what was fun now sometimes meant trouble later. Even my essays gave the "impression of being hastily conceived" and contained "a great many errors in the mechanics of writing," displaying my regrettable taste for "sensational titles such as *Ku Klux Klan* and *Murder, Incorporated*."

By this time, I had fallen in with a group of restless boys in ties and jackets. I was experimenting with new curse words. I found a silver flask in one of my father's drawers and filled it with whiskey, taking it to school to hide in my locker and reveal to a select few. In the woods above the Potomac Parkway, I buried a jar filled with cigarettes, and one day John Brinkley and I smoked till we got dizzy and decided to throw eggs at passing cars, which was so much fun we raided the fridge and unloaded a whole carton. Then I ran down to the shoulder and slopped a bottle of ketchup across the side of some car and we ran and ran and ran and ran—until the police caught us and took us home.

In my next report card, Ricker nailed me. "John has entered early into the unstable and troubled period of adolescence, and he has not yet found himself." At the bottom, a headmaster named William H. Triplett added a single handwritten sentence: "Unless John's language, conduct and attitude improve between now and June, his invitation to return to Landon will be withdrawn."

That kicked off the summer of 1967, which began with huge anti-war protests in New York and San Francisco. By then my father's worst predictions about Vietnam were coming true. We were, in fact, stumbling from one ineffectual leader to another and the Vietcong were pouring across those long indefensible borders. My mother remembered him saying "Bomb the dikes," which suggests that he believed we should take the fight to the North. He was horrified at the idea of abandoning the South Vietnamese. But she also remembered him saying that the war was a terrible mistake. At the same time, he was disgusted by the younger generation, with its simplistic solutions and vulgar behavior and total lack of respect. His generation may have protested, but they conducted themselves with some basic dignity and decorum. They didn't dress in rags and thumb their noses at their parents.

Perhaps this is one reason he decided to address my discipline problem by giving me some adult responsibilities. These came in the form of the red Toro lawnmower in our garage. Soon I knew every bump in the backyard, especially the big knob of quartz that clanged the blades when I got careless, which was often. Then I got a brilliant idea: why not drive a stake into the middle of the lawn and tie a rope from the stake to the lawnmower and set the speed lever on the little rabbit? Then the lawnmower would go around in circles, thus wrapping the rope around the stake and making the circles smaller and smaller—as I sat on the patio in the shade.

When my father saw what I was up to, he started to sputter. "You're headed for trouble! *You're always looking for the easy way out!*"

Not long after that, I made the mistake of talking back to the old man one time too many. He dragged me to the front door and gave me an actual physical boot in the ass. I remember tramping in the weeds along the Dolly Madison Highway, the outrage pressing my ribs so hard I could barely breathe, swearing a solemn oath that I'd show him—go to Harvard, become a big shot, do something important. At some point, I stopped at a church to linger in an anteroom, dreamy with the dark shadows and colored windows. When they noticed me, I moved on to a colonial restaurant and curled up in a wooden wagon on the lawn, staring up at the stars and wallowing in the drama of my solitude. But it got cold. After a while I wandered up to McLean and found an open laundromat, which was warmer. But I felt exposed in the blue light. A police car drove past and slowed down to look at me. Just when I had given up and started walking home, it came back around and the police drove me the rest of the way. I remember my mother standing at the door, shaking her head.

Now things are sputtering to life. I'm at Mrs. Shippin's dancing school, wearing white gloves and dancing with the girls in stiff dresses. It's like we're wearing doll clothes, marching in and out of the cuckoo clock.

I'm leafing through an old history book in the library, mesmerized by a color plate of a Roman slavegirl with her wrists bound to marble columns, her breasts bare.

I'm sitting on the floor of my room, studying my new Simon & Garfunkel album. What the heck is a desultory philippic?

My teachers are constantly sending me to the headmaster's office, where I've spent many hours burning in humiliation, trying to hold back defiant tears. Mr. Ford is particularly dismayed by my performance in English. "His understanding and incisiveness are very often exceptional for one his age, but sometimes this inquisitive attitude leads to argument for its own sake. Disrespect and a whining tone will often result."

That fall, Landon's headmaster sends my parents a letter requesting a face-to-face meeting. My father sits at his desk slashing angry black lines under the most damning phrases: ". . . his recent quiz grades have been very low, and they are indicative of superficial, if any, preparation. His homework assignments, when completed, have been done with little thought to neatness or accuracy, and I fear that if this trend continues, he is in for a rather difficult time."

The irony is that just as I'm hitting the wild years, my father's whole life has become an opera of scrupulousness. He's *Director of Training*, for chrissake. He's Chairman of the Committee for Professional Manpower and a member of the General Counsel Inspector General Committee and the Agency's representative to the Federal Executive Institute. He spends his days writing memos like this:

For over a year the Clandestine Services have carefully analyzed every audio operation where compromises have occurred in an effort to determine probable causes. The DD/P is interested in having some of these studies used by OTR and TSD in training CTs and other operations personnel. The matter is now being studied.

His biggest concern is that people in Training aren't taken seriously by the rest of the agency. Spies who get rotated to him take it like a criticism. People think they can't cut it in the field. Fighting for the best agents, he sometimes pushes too hard. This leads to "a good deal of criticism," one of his friends tells me later. He's second-guessing the brass and ranking his division's priorities too high.

He's in one of his stubborn moods again.

This may be the time he sings a little rebellious light verse at one of the Agency parties. Modeled on a Gilbert and Sullivan song but unsigned to protect the guilty, it's called *A Simple Tar's Story*.

O sailors all, where'er you may be,
If you want to rise to the top of the tree,
If your soul isn't fettered to a quarterdeck stool,
Be careful to be guided by this golden rule.
Chorus: Keep your minds a perfect blank
And remain at sea
And you will be directors
Of the Agency.

As it goes on, it gets downright subversive.

Of intelligence I had so little grip,
That they offered me the directorship.
With my brass-bound head of oak so stout,
I don't have to know what it's all about.
Chorus: He pleased the commodore
So mightily
That now he's a director
Of the Agency.

That summer, the summer of 1968, he tries to address my distress by taking me on a fishing trip. Like the trip to the Blue Ridge Mountains, this trip stays in my mind in a series of snow-globe memories:

26. We drive north into a night so foggy our headlights blind us and we end up throwing our tent up in a field, and in the morning we pack up in the rain and leave behind a flat patch of grass.

27. He points out the "quaint little churches" of New England. There are long stretches of silence.

28. We stop at a lake and I sit with him in the rain in a musty three-sided cabin, losing myself in George Orwell's *1984*. The next day the sun clears and there's a girl and I spend the whole day on the lake with her.

29. In Maine, we camp on a logging road, a red scar cut into a sea of pine trees. In the morning, he teaches me to drink instant coffee with lots of sugar and Cremora.

30. Crossing into Canada, we begin counting the Irving gas stations, making a game of it, starting to relax a little bit. It's an odd feeling, being silly with my father. I'm not quite sure but I think it may be fun.

31. At a lake where I pull sunfish out of the warm shallows, he refuses to fish with worms. That's not real fishing, he says. My sunfish become pallid things—but I keep on fishing.

32. As we get close to home, we cross a bridge in the rain and I see the rusted bolts and the bottle-green river and suddenly the songs on the radio come through with vivid clarity: *I Think We're Alone Now; (Everyone Knows It's) Windy; The Rain, The Park and Other Things*. Somehow I know I will always remember them.

When we get back, the old momentum resumes. By 1969, my sister and I are both spinning out of control. She's a high-school junior, playing the nympho in *No Exit* and going to see the daring French movie *Belle de jour* and writing in her diary about "drinking, dancing, kissing, love, talking, fun & music." I'm in ninth grade, famous at school for writing a pornographic story that involves a stewardess and a hijacking and making fresh progress in my criminal career—after reading an article about shoplifting, I lead my buddy John into an electronics store and take a tape recorder off the shelf, pull off the price tag and march straight up to the counter to tell the clerk that I can't get it working. And sure enough, he puts batteries in it and sends us out the door. The rest of the day feels legendary, walking along the railroad tracks and wasting time and somehow ending up at a party with a rock band playing so loud it feels dangerous.

Then I start to explore my father's wet bar. In my memory, this glamorous little room is a significant detail of our house, a hidden alcove off the study with mirrors and bottles and a half-sized refrigerator. Along with his big TEAC tape recorder and long-playing dance tapes, it's one of his attempts to grasp the good life. At his parties, he mixes martinis in a silver shaker and puts them on a silver tray for me to serve. "Take this to Mr. Helms, and be careful because he is a very important man."

Perhaps this is the night mentioned in one of those declassified CIA memos: "On Saturday night at Jocko's the Director and the undersigned were talking for a while after dinner regarding youth today . . . Mr. Helms's quick reaction was to suggest I take off some time and go recruit the bright, alienated youth."

I like to think this is the moment when I bring Mr. Helms his martini.

One night when my parents are out somewhere, my friends and I decide to have a drinking contest. We pour gin and vodka and *crème de menthe* into shot glass after shot glass.

I win.

That winter, my parents seem to be fighting a lot. My mother is in her early forties and having a depressed husband in his fifties must be a strain for her. Sometimes she argues with him, telling him he's too hard on me. One night, my sister overhears her saying something about divorce.

I try to tune them out, spending most of my time in my room or with friends or riding my bike alone. I'm drawn to construction sites, where the raw earth and unfinished rooms trigger something in me. One day, I step on a piece of glass and cut a gaping hole in my foot. When I get home the house is empty, so I wash up in the tub and wrap my foot in towels, wiping away all the blood so that nobody will know.

That earns me two weeks home from school with my foot on a pillow, which I spend reading a book called *Really the Blues*. It's about a white kid who plays jazz in brothels full of gangsters and "vipers." I love it from the first lines:

> "Music school? Are you kidding? I learned to play the sax
> in Pontiac Reformatory."

I convince my parents to buy me an electric guitar.

Then it's spring. Down in Georgetown, new shops that seem to have sprung up overnight burn incense and blast rock music and have weird color-ful posters of bands with goofy names like Jefferson Airplane. Walking along the sidewalk one day, we pass a hippie in engineer's pants who calls out in a loud voice: "Acid!" A few days after that, Chris and I are watching his older brother paint Blue Meanies on the walls of his room. When he's finished, he pulls a plastic baggie out of his pocket. "You guys ever smoke grass?"

I don't even hesitate. When I get home, I feel everything coming into crisp focus. I notice my peripheral vision and especially my peripheral *thoughts*. For the first time, I see all the colors in my skin. For the first time, I hear the songs on the radio:

Those were the days, my friend. We thought they'd never end . . .

One weekend, I tell my parents I'm going to a friend's house and pedal down Dolly Madison to the canal that flows all the way to Georgetown. Five miles later, I walk up Pennsylvania Avenue and find the hippie in engineer's pants. He sells me something called kif and on my way home, I hide in the bushes just off the footpath and smoke it—and float home along the dreamy gliding water.

A few months later, I buy the *White Album,* close the door to my room and listen with my eyes closed. I study the song titles, the photographs, the sliced apple on the label. Then I ride back down the canal and find the guy in the engineer's pants. But this time he's out of kif.

"Let me have some acid then."

"You?"

"I can handle it. I've done it before."

When I get home, my mother asks me to come talk to her. So I sit in the swan-armed rocking chair babbling about some bland thing, school or the weather or Mrs. Banfield's azalea garden, and my lips seem to be mov-ing normally even though the hair on my arm is growing at an alarming rate. After a while, she seems satisfied and lets me go, and I realize she doesn't even see me. I can get away with anything.

Around this time, my father asks me to join him in the study. So I sit down in one of the red leather chairs and he sits down in the other, tapping his cigarette into the crystal ashtray on the Kittinger table. He's been posted to Korea, he says. We will be moving there in June. He wants to give me the option of staying at Landon as a boarding student and he wants me to make an informed decision. "You've reached the age," he says, "when you're old enough to know what it is I do."

Beyond that, I don't remember his exact words. My biggest concern is being left behind, and I'm hurt that he would even suggest it. "I want to go with the family," I say.

And that's it. The conversation is over. He doesn't explain why he wants to go so much that he's willing to give up his new house or our chance to put down roots in America or his last shot at a top job in the CIA. I think he says that Korea is a fascinating country where I will have many rich experiences, but he doesn't explain that North Korea is the most totalitarian country in the world or that Dick Helms is giving him the post as a reward for his hard service in Vietnam or that it's his last job before retirement, which by CIA rules is only four years away. I was probably too much the oblivious teenager to listen anyway. But over the next few days, I start to ponder the CIA part. From TV and *Time* magazine I have the vague impression that people think the CIA is a bad thing. But it's also the era of James Bond and *In Like Flint,* the groovy James Coburn movie where he fights off an international conspiracy of Playmates trying to conquer the world. Neither image fits my somber father so I go poking for clues in his dresser drawers and, lo and behold, hidden under the perfectly folded handkerchiefs and boxer shorts, I find a gun. And not just any gun, but a snub-nosed .38 with a gleaming oilslick barrel and his initials—*our* initials—carved into the ivory handle. Next time he's out of the house, I show it to one of my friends and we take it out into the woods. I aim it at a tree and pull the trigger and the crack of the gunshot explodes so loud across the hills we take off running and keep running until I get the damn thing back in the drawer.

Next time I look for it, it's gone.

This feels right, since there's an unreality to this whole idea of my father being a spy. He wears a suit and horn-rimmed glasses and goes to the office every day. He reads a lot and worries about the ivy on the hill. He's the same Dad he always was. But I keep trying to tease my little tendril of hope into bloom. One night, a man comes to stay overnight and when he and my parents leave for the day, I snoop through his luggage and find a reel-to-reel tape. Carefully peeling back the Scotch tape that holds the box closed, I take it downstairs and put it on my father's giant TEAC. It's something about Kennedy and foreign policy, nothing that makes any sense to me, but I'm meticulous about replacing the tape and repacking his suitcase exactly as I found it—tradecraft, Mr. Bond.

Then the family gets ready to break up. Jennifer's heading to Indonesia to spend the summer with Uncle Dean, I'm getting ready to go on a canoe trip with my old Boy Scout troop, Dad's heading off to Korea and Mom's staying to supervise the packing. By this time, Jennifer has a bad case of what psychologists call "boundary issues." The week before she leaves, she cuts school to buy beer with a couple of boys named Sprague and Bob, detailing the results in her diary: "We ended up drinking two six packs on a No Trespassing road. I was drunk and Bob & I kissed a couple times while I was in his lap, then Sprague & I went back to school and he kissed me too."

I'm worse. One night, I tell my parents I'm going out camping and go down to Georgetown, where I meet a guy on the street and "crash" on the floor of his apartment. In the middle of the night, he crawls under my blanket. Too scared to participate and too lost to refuse, I pretend to be asleep. A few days after that, one of my friends says something about niggers coming up from Washington and I wait until it's dark and slip through the woods and hurl a rock through his bedroom window: *I am a nigger from Washington!*

Before my father leaves, his CIA buddies throw him a goodbye party. After they down a few martinis, one of them gets up to read a poem:

Jocko was "dis-Lodged" from hot Saigon
In training he did cool.
Now thawed and rested he's off again.
To thrive in the heat of Seoul.

It goes on for another twelve stanzas, full of jokes about the halls of old Blue U.

Give me a raw lad
Witty and spry
And eleven months later
I'll give you a spy.

Three weeks later, I'm lying on the deck of an ocean liner, reading *Atlas Shrugged* and making a pact with myself to read all the books in the world.

32

•

SEOUL, 1969

Waiting in Seoul in a temporary apartment, my father plunges into his work and finds himself overwhelmed at how much there is to do. But he takes the time to write us a few of his oddly impersonal letters. "At first things in Korea may seem strange and you may feel uncomfortable," he warns, "but within a month you'll probably feel like an old-timer."

> Remember that the drivers and the maids speak and understand English, so be careful what you say. When people ask you what your father does in the embassy, just say that I'm a "Special Assistant to the Ambassador."

But when my sister writes to him from Jakarta to pass on her observations about the poverty in Indonesia and the feckless embassy brats who spend most of their time drinking and smoking pot, he responds with three handwritten pages that strike quite a different note.

I'm overwhelmed at receiving your last long letter. I'm getting homesick too, not for a place or a neighborhood, but to
see the family together again.

Her tone is so adult, he says, she seems to have grown up overnight.
But he hopes she won't focus too much on the negative because every
culture has its art and beauty and we should try to look at it with a sympathetic eye (while not, of course, ignoring unpleasant realities). And hopefully she will find Seoul more to her taste.

The Seoulites are people in a hurry. They run across the
street instead of walking. They're clean, good-looking, mus-
cular, well-dressed. No one lies in the street here—they'd be
run over.

He mentions this, he says, because Americans overseas are so often
criticized for living with other Americans in a "Golden Ghetto." He hopes
that will not happen to us.

And yes, there is pot here too. It's not illegal but the embassy forbids its
use as a matter of official policy.

If an American teenager becomes a pot problem case, the
whole family is sent home, including the father. But most of
the kids don't use it here any more than they do at home. At
least, I hope not.

Two weeks later, my mother and I arrive and Dad takes us on a tour of
the city in his black diplomatic car. With our new chauffeur impassive
behind the wheel, his broad face and mirrored sunglasses evoking our
chauffeurs from Vietnam and the Philippines, Dad sketches out the political situation. Two years ago a group of North Korean guerillas slipped
across the border to attack the Blue House, the Korean equivalent of our
White House. They made it as far as those woods right over there, just to
the left of our new home. And perhaps we'd heard about last year's *Pueblo*
incident, when North Korean soldiers attacked a United States spy ship
and kept the survivors in prison for eleven months of torture and interrogation. And the DMZ—that's short for demilitarized zone, son—is just
twenty-six miles away, with hundreds of thousands of North Korean troops

drawing a hostile line between North and South. So things are tense and we must present ourselves as ambassadors of our country at all times. We can never act like "ugly Americans."

And remember, many Koreans can understand English.

We pass masses of electric signs in Chinese and Hangul and little square taxis and big stinky buses, men holding hands with men and women holding hands with women and so much bustle and movement it all seems vaguely unreal, like something out of a documentary film. Perhaps because of the long separation, I take my father in too. With his dark suit and tie, his horn-rimmed glasses and bald head and formal manner, he seems to come from a different world.

In the grassy divide between lanes of traffic, an old woman squats to pee. Dad doesn't seem to notice.

Then we pull up to a Military Police hut and the MP waves us through, and suddenly big white stones line the road. The trunks of the trees are painted white too, and there's so much wooded land it looks like a wilderness park. This is Yongsan Army Garrison, the headquarters of the Eighth Army. And here's the PX and the library and the hospital and those are the barracks. Then there's another gate and a tree-lined avenue and another gate leading to the lower half of the base, where sergeants and officers raise their families in hundreds of identical cinderblock ranch houses.

And there's the high school, also cinderblock. And the teen club, cinderblock.

Back on local streets, we drive past Coulter's Statue and Itaewon, the U.N. housing and Yonsai University, where student protests periodically explode into anarchy. The Korean authorities do not take protests lightly, my father warns me. After the demonstrations in 1964, the president imposed martial law. Kids who try to grow their hair long are marched to the police station and given a buzz-cut.

"Remember, son," my father says, "don't be an *ugly* American."

But I seem to keep saying the wrong thing. At somebody's welcome-to-Korea dinner I talk about the fabulous Kobe beef in Japan and how they feed the steers beer and massage them to make the meat soft, and later my parents are angry with me. How could I praise Kobe beef to the hostess when *she* was serving beef?

And my mother is absolutely disgusted with our new house. Embassy Compound Two is surrounded by a wall deep in the heart of the city, a few dozen houses and maybe twenty apartment buildings, plus a store and a pool and a tennis court and a Marine barracks for the embassy guards. The Blue House is down the street, a Buddhist temple next door, and behind us there's a tangle of alleys and curved roofs that climb up a series of hills. But the people who lived here before us had dogs and a cockatoo and Mom gets flea bites just sitting in the living room, so once again she has to get rid of every rug and piece of furniture and order the ritual Richardson fumigation. Then she sets the domestic staff—we have a cook and three helpers—to work. We've barely finished that when Dad invites the Korean prime minister for dinner.

Fortunately, an embassy wife named Cherry Murray whips up an elaborate Peking duck to save the day. "I hope you play bridge," she tells my mother.

This time she learns. And she joins the Bamboo Circle and the Garden Club and signs up for painting lessons and works on the Children's Hospital Relief Committee and takes notes on Korean culture. Perhaps she's trying to keep up with my father, who comes home late every night and often goes right back out to Korean drinking parties, where tradition and diplomacy require him to spend hour after hour throwing back shot glasses and yelling *gombai* while geishas refill his glass.

And the engraved invitations keep pouring in, on stiff cards with embossed crowns and gold borders: Her Britannic Majesty's Ambassador and Mrs. Jeffrey Peterson request the pleasure of your company. Ditto the German ambassador and the Israeli ambassador and the ambassadors of Thailand and Australia and the Netherlands and Mrs. Verkade and General and Mrs. J. H. Michaelis and the Minister of National Defense and Mr. Lee Byong Doo of the Korean Central Intelligence Agency, all requesting the pleasure of their company.

Left to myself, convinced that I am some kind of budding alienated genius, I write reams of horrible poetry:

> I try to be a fake; hide, my mask illusion
> Being part of people just brings on more confusion.

But soon we go to dinner at another CIA house on the compound and I meet a couple of teenagers with promisingly fuzzy hair. Tom's my age,

Bob a year older. Bob's got that cynical smirky style I recognize as cool and before the night is out, I confess that I smoke pot. Within a day or two, they take me up a steep little street to a place called Itaewon. Now here's an alternate reality: GIs lurching around with red eyes, Korean bar girls in miniskirts, little stores selling tie-dyed T-shirts and satin jackets with maps of Korea stitched onto the backs. We pass the Lucky Hotel and the 007 Club. The bar girls giggle and promise us "number one long time."

We duck into a dusty little store and Bob tells the mamasan we want happysmoke. She brings out a rumpled cigarette pack. "Number one happysmoke, five hundred won," she says. For a buck fifty, we get twenty cigarettes that have been carefully emptied and refilled, the tips twisted into little points.

Looking back, it's astonishing how quickly things begin to fall apart. Late in August, my sister arrives from Indonesia on a gust of manic energy. She's tanned and voluptuous and genuinely stunning, the kind of girl who makes men run their cars off the road. My father breaks out the champagne and she tells us about the dances and the pool and Jakarta and Bali and most of all, about a guy named Mike Short. Not only is he in love with her, he's coming to visit in a few weeks. And she's thrilled to see us and remembers all the little presents Mom hid in her bag and even the stuffed doggie I gave her at the airport.

But the fights start the very next day. Dad is giving a party and she wants to wear her sexy new dress, which Dad thinks is horribly inappropriate. In her diary, my sister rages: "I will not be me as dictated by my parents! I will be me as me, goddamn it!" When Dad offers to get her a tailored winter coat, she insists on a floor-length bright purple coat with a hot pink lining. And that weekend she meets a marine named Jim and she and Dad start fighting about that—she's too young to date marines, especially the marines that guard the embassy. He has to work with the men.

Then school starts and they send home all the boys who grew their hair long over the summer. At dinner that night, sitting around the gleaming table as the servants pour soup, my sister and I try to convince our parents that the Army is being ridiculous. But Dad gives us a stern lecture on "running with the herd."

A week or two later, my mother goes off to Tokyo. Maybe they've been fighting, maybe she just wants to buy furniture. But in her absence there are signs that Dad is having trouble. Every so often, he goes off with his CIA interpreter and comes home so drunk his driver has to help carry him up the stairs. One night I help and when we get to his room, he sits down on the edge of his bed and gives me a stricken look I will never forget. "I had a brother, did you know that?"

"You did?"

And he tells me about the uncle I never knew, a decent young man who liked airplanes and hadn't quite found his way in life. I'm sitting on the stool next to his bed looking up at him. He's half slumped, his body gone loose inside his suit. He says he was at Berkeley when he got a phone call from his cousin saying that Leonard was dead. He rushed down to Whittier but it was too late. "Everyone said it was an accident," he says, "but how do you shoot yourself *acc-i-dent-ally* with a shotgun?"

It's the first time I have ever seen him cry.

My sister has the same experience and records the details in her diary. Stop worrying about virginity, he tells her. It's not important. Just don't get pregnant or get a venereal disease. Try to be wise and understanding and give affection and love, because love is the most important thing.

> He had been drinking & was open & I saw him, his hidden real human self. He's sad for his family, mother, dad, brother—shot himself at 20. Also Vietnam and war. Wept. I love my father.

I wish I could say that I react to all this with a deeper sense of wisdom and understanding. Instead, I discover makoli, the milky fermented rice wine cherished by Korean peasants. On September 15, Jennifer makes a note: "Jerk went drinking." Two weeks later, Dad confronts her: "Why was Jocko bombed?" Then Jennifer meets a dashing Tennessee wild boy named Rusty Riggs and he teaches me about the wonders of my father's medicine cabinet, which turns out to be full of fat red Seconal sleeping pills, extra-strong because Dad's been taking them ever since the end of World War II and still drinks dozens of cups of coffee a day. Three of them put me in a hot bath of woozy pleasure and inspire me to new heights of defiance. Again, the evidence is in my sister's diary:

> Jocko and Dad got into a fite over a hair cut and Jocko
> walked out. Dad going to have a heart attack & Jocko going
> to have his damn principles. What now?

And everything seems to be leading in the same direction. One day on the schoolbus, a kid with big horn rims hands me a copy of *Cat's Cradle*, which is about a substance called Ice 9 that ends the world in a mad-scientist technocrat bang. By this time Tom and Bob have introduced me to the gnostic study of record bins at the PX, where we select records based on rumor and cover art and I discover that "(c) Ice 9" appears after every Grateful Dead song. And one day in the Army library I'm researching a paper and I come across a book called *Modern French Theatre*, which attracts me because there's a naked woman on the cover. Inside are plays by Ionesco, Alfred Jarry and Antonin Artaud, one episode of anarchy, surrealism and depraved sex after another. It all fits together, helping me shrug off the absurd world.

My sister's even worse. She's discovered something she calls "corn-flakes," which help her get through her term paper on the Nixon administration.

> They make you concentrate and do everything fast and
> you don't want to eat and are very thirsty and can't sleep. It's
> now 5:30 and I'm still not tired. Will continue to take a mod-
> est amount for weight.

At this point, her "weight problem" is all in her mind. But she wants to look like a ballet dancer, so Mom and Dad agree to give her a dollar a pound.

One night, she announces that she's getting married to Rusty. At first, Dad is amazingly patient about it, telling her there's no such thing as falling in love. If it's real and Rusty can support her, they'll grow into love.

Then she says she's decided to become a professional dancer and again he's patient, telling her that it's a very risky profession and she should go to college first so she has something to fall back on.

Another night, he asks us not to smoke pot on the logical grounds that the science hasn't yet been done to show what the dangers are. With alcohol and cigarettes, at least you know the dangers. And please remember that the most important thing is to be kind, to listen to other people. "I never learned anything with my mouth open," he says.

But after my mother catches Rusty trying to sneak in the window, his first reaction is one of his scary rages, telling Jennifer she's going to end up a *chorus girl* which is *just above prostitute*. Then he catches himself and apologizes, telling her about the Italian baroness who first seduced him, a woman with a very natural approach to life. "She used to say, 'anything animals can do, we can do.'"

For Jennifer, this comes as quite a surprise. But it's not completely out of character. With his skepticism about religion and his quotes from Jefferson and Voltaire, he's always been a bit of a freethinker.

"Sex is natural," he says.

"So you don't think I have to be a virgin when I get married?"

"Actually, Jennifer, I've always considered the obsession people have with virginity a perverted standard from the past. A healthy man would rather his wife be experienced. Sex should be sensual. One should always satisfy one's partner."

That night she writes in her diary: "Put sex in a different lite. Grateful to him. Love him very much."

But she keeps pushing his limits. When he takes us out to dinner with a Korean police chief, she takes some "cornflakes" in her purse. A month later, she takes so many cornflakes that she passes out at school and has to stay in bed for two days. A few weeks after that, she's decided to marry a GI named Fred—and Dad is raging again, saying that he won't give her permission and if she does, in fact, decide to run off with him then Fred can damn well pay for her college expenses. Another week and she's presenting a detailed plan to graduate early so she can join Fred in New York. Mr. McGuffin has already agreed to give her credit for her performance in *Guys and Dolls*. Dad tells her she's immature, she falls in love every three weeks, she's not even a good dancer and she'll be lucky to end up a chorus girl—which is *one step above a prostitute*.

The rest of that winter, our lives seem to zigzag like a lurching drunk. One week we're seeing The Love Bug, the next week it's Alice's Restaurant. One day Fred is giving Jennifer the purple heart he won in Vietnam, a month later Jeff is telling her his ambition in life is to be a failure who sits on the beach and writes poems. And when Jennifer ends high school on the honor roll, Dad asks her not to leave until May because he's afraid that dancing is such a tough profession. Somehow it turns into yet another "fite" about bumping and grinding and the chances that she'll end up a prostitute. At bedtime she writes in her diary. "I know he loves me and is trying to protect me, but he can sometimes be very cruel."

Then it's my turn. One night a bunch of kids and GIs take over a whole floor of the Lucky Hotel and the party spills out of the rooms into the halls, wall-to-wall grunts and freaks and bar girls, and somewhere in the haze I meet a Korean girl who dresses like a hippie and speaks a little English. Jennifer takes note: "Jocko didn't come home tonight. Took out Korean girl—Blondie. Probably spent the night in Larry's hooch—the 4 of them. 'Rents worried."

I remember a dismal little apartment building in Itaewon and a tiny room heated by a charcoal stove. Blondie uses tongs to change the round bricks. She cooks me rice and eggs. She introduces me to her mamasan.

No wonder school seems increasingly silly. When I'm late and a teacher asks me for a note, I get Tom to write one and sign it with his real name. The teacher throws me out for being flippant and I just laugh. And Mom and I have a fight about my hair and suspected drug use and habit of interrupting and Jennifer consoles me with a long talk about sex and life. We're co-conspirators, allies against the world:

> Mom and Dad were very upset cause Jocko got caught
> drunk & gave the MP a lot of backtalk. They can't talk to him
> cause everyone walks out. I'm the only one who can talk to
> him or understand him.

Around this time, Si Bourgin comes out for a visit. It's been a few years since they saw each other and at first, he finds my father "sweet and solicitous" as always. But then they start talking politics and Si makes the mistake of criticizing Spiro Agnew, and Dad jolts forward so suddenly it's as if he's been hit by an electric current. "*What!?* Do you mean to say that's what you think of Spiro Agnew?"

At this point, Agnew is still years away from the indictment that led to his disgrace and resignation. But with his florid attacks on left wingers and war protesters and other "nattering nabobs of negativism," he's become a symbol of the right and a lightning rod for critics of President Richard Nixon.

Startled, Si blurts out the truth. In fact, he's probably been more generous than he really feels. And he's stunned when his host and old friend explodes into a tirade about the left wing and the know-nothing zealots who are tearing down the country with their simplistic solutions to *extremely complex geopolitical problems.*

Mom keeps saying, "John, what's the matter with you? Si's our *friend.*"

But he won't stop. He's in a rage.

In the morning, he calls to apologize. But for Si, it's a lasting shock, "one of those seminal moments where you discover how deeply somebody feels about something." In a few minutes, he's glimpsed how far apart they've traveled over the years, how conservative my father has become. Musing about it much later, he offers me his sympathy: "When I think of John during that period, he would be a very hard father—inflexible, speaking with authority, and expecting his children to jump."

But what strikes me about the story is my father's sudden lurch from sweetness to anger to apology, which seems a perfect symbol for these years. Over and over, he *tries* to be understanding and permissive. Lulled by Shostakovich's Fifth Symphony one night, he has a warm talk with Jennifer about "the struggle of life." Or he walks home from work in the snow so she can take the car to a rehearsal. At the Officers' Club for dinner, he lets her order a double martini. And he's so proud and happy on the opening night of *The Taming of the Shrew,* telling everyone he's a bit disturbed at what a convincing Kate his baby girl turned out to be.

But he worries. Maybe her voice is pitched too high, maybe it will hurt her career, maybe she'll end up broke and miserable and alone. And don't forget, showbiz is a twenty-four-a-day job.

Now that I've announced I want to be a writer, he does the same with me. How are you going to make a living? How do you know you have the talent? What themes are you going to choose? How is your work going to fit into the history of Western literature?

Then he goes off to Hong Kong on a three-week business trip and we shrug off his anxiety with nights as wild as we can make them, me at Blondie's hooch and the 007 Club and Jennifer at the Officers' Club, talking a waiter into putting martinis in a water glass and drinking toast after toast to love love *love* and dancing with a GI who wants to marry her and coming home drunk, where she tells Mom not to worry because everything's going to be all right, she has talent and so do I and we're *going to be okay.*

Then Dad comes back.

But she's hurt because he brought books for me and not for her. And she had just read *The Godfather.* "I have become determined to develop my mind," she wrote in her diary. "Dad's going to help me now. I love him. 1st job—read *Taipan.*"

And next thing you know, there's another lurch and they're getting into another big fight about the latest GI and this time Mom takes over and for the first and last time, the two of them reach the point of "slapping, kicking, name-calling etc." And Mom keeps after her until she promises to dress and act like a lady, with manners.

But a week later, she gets accepted by a small private college in New York and decides to celebrate by ending her "present virgin state of physical being." After spending a last day drinking with the longhairs, she packs a small suitcase and heads off to the Tower Hotel, where she and the GI-du-jour drink a couple of double martinis and eat some chocolate. Then she changes into a bathrobe and he puts on a condom.

When it's over, she looks at the clock: 8:05.

But sex doesn't settle her down. Instead she gets so wild, I fade into the background. She flirts and dances with other men until her new boyfriend loses his temper right there at the Officers' Club in front of the whole family, which is probably exactly what she wanted him to do. One night at a dance at our house, she attacks an American general so shamelessly—at one point, she nibbles his ear on the dance floor—that he ends up sliding his hand down over her ass while his wife watches from the sidelines, getting so furious that she ends up taking my mother aside and insisting that Jennifer write a formal letter of apology.

The next day, Mother and Jennifer have an epic fight. Jennifer says she's just seventeen and the general is a grown man, let *him* write the letter of apology! But Mother keeps at her until she gives in. A few days later, Jennifer makes her debut in *Gypsy,* wearing a nude bodystocking with ostrich feathers and working electric lights. From the seats, it looks like she's naked. She makes her big dance number as sexy as possible, a real bump and grind. And after the show she marches into the Officers' Club like a star, tossing her bright purple coat over her shoulder so the hot pink lining falls open with a flourish.

But Dad seems pleased. "Not bad at all," he says. "She may make it."

At this point in my life, I'm doing my best to ignore the world of adults. I have the vague idea that Korea is ruled by an "authoritarian" leader named Park Chung Hee and that Korean kids are afraid of the cops. I have the vague idea that back home the hippies are protesting the war. Eventually I learn that Park is another Ngo Dinh Diem, suspicious of foreigners and increasingly convinced that the only way to build his country into a modern power is by dictatorial control of the economy and everything else. I learn that once again, my father is being accused of downplaying negative reports. I learn that my father's counterpart at the Korean CIA is a brutal man named Lee Hu Rak whose agents will soon become notorious for torturing opposition leaders and protesters.

But that knowledge comes much later. In March, when some Japanese Communists hijack a plane with samurai swords and try to fly it into North Korea, I get a kid's-eye glimpse of my father in action. Sitting in the library, switching between the phone to Washington and the portable direct-receiver walkie-talkie that connects him to the Blue House, he helps engineer a brilliant diversion. They route the plane to Kimpo Airport—our airport, right here in Seoul—and the South Koreans scramble out a new sign disguising it as a North Korean airport. The plane lands and they trap it on the runway.

Then the hijackers threaten to start killing the passengers.

Dad goes back to the phone and the walkie-talkie. He's somber, focused, intense. This is what he lives for, the fight against totalitarianism made tangible. During a lull, he tells us that if they let the plane go to North Korea the passengers will certainly be tortured.

Then he goes back to juggling calls and Mom herds us upstairs to give him some privacy and maybe because of the tension or because I said something snotty, she starts nagging me about my long hair and suspected drug use and my habit of interrupting my father, which shows no respect at all.

Don't bug me, that's my motto.

Three days later, when the hijackers release the passengers and South Korea lets the plane go, I don't even notice.

That spring, the spring of 1970, the United States invades Cambodia and American students explode in mass protests. At Kent State University in Ohio, they burn down the ROTC building and the governor sends in the National Guard. On May 4, the guardsmen fire on the students and kill four of them.

None of this makes much of an impression on me. It gets a little play on the Armed Forces Network, but I'm more interested in *Mission Impossible* and *Bonanza.* Then Mom announces that she's taking us to see the International Exposition in Kyoto, sort of a goodbye trip for my sister, and to me it seems like an excellent opportunity to smuggle drugs with my black diplomatic passport. With the black passport, I can't be searched or arrested. I'm untouchable. So I go to a GI ville outside of Seoul and wander through the mud alleys and wooden houses until I find a mamasan who offers me a cup of barley tea and sends her boy out to get a huge Santa bag full of raw stalks, which I happily sling over my shoulder and carry home on the train. After the servants are asleep, Jennifer and I go downstairs and spread it out on tin foil and dry it out in the oven. Man, it reeks!

By now everything about my mother irks and embarrasses me, from her bouffant hair to her airline stewardess pantsuits to her obsession with proper behavior. So when we get to Japan, I spend most of the trip wandering around alone. I find my way to the red-light district and unload a few bags of pot, write lots of poems, munch my sister's cornflakes. While my mother and sister sleep in, I go to the Expo by myself. There are moving sidewalks! A picture of Che Guevara! Buildings like spaceships!

One day I get back and find Mom in a tizzy. "Do you know what I found in your suitcase?" she asks. "Pounds and pounds of pot!"

She threatens to flush it down the toilet. But I point out that this is a bit impractical and somehow manage to convince her to let me get rid of it instead. So Jennifer and I grab a cab to the red-light district, where we give the whole bag to some startled hippies.

The summer continues like that. When we get back to Seoul, Dad takes us on Sunday cultural excursions and to movies. Once he takes us for a weekend at the lake house of his golfing buddy, a bigshot Korean industrialist known as Dynamite Kim. But we just go through polite-child motions and rush back to our private obsessions—Jennifer to her GIs and me to music and camping and drugs and most of all to my new girlfriend June, who has long dark hair and olive skin and happy smiling eyes, who loves Frank Zappa and *The Teachings of Don Juan* and *The Tale of Genji*, who kisses me on the riverbank in the moonlight and spontaneously peels off her shirt. So we lie to our parents and steal our first night together, renting a flat-bottomed picnic boat and paying an old boatman to anchor us out on the river.

But in the middle of the night, the anchor pulls loose and we go drifting helplessly down the river, which turns out to be a perfect metaphor for the entire year. When Jennifer leaves for college on a grand tour that includes stops in Israel, Greece, Italy and Spain our letters show us all flailing at reeds. In my scrawled notes, I tell her not to use drugs too often, to listen to Elmore James and Big Mama Thornton, to read *Main Street*. I rhapsodize about June—"her eyes remind me of tadpoles"—and tell her about the gorgeous campsite I found out in the country, right next to a crystal stream with mountains sloping up on both sides.

> I'm going tomorrow with five or six friends and ten tabs of
> acid. Mother has busted me about four times since you left. I
> don't think she cares anymore.

Mom sends cheery letters with lists of things to do and updates on finances and schedules. She chatters on about luncheons and party arrangements. And every now and then, she lets a glint of irritation show through: someone asked about you so I mentioned the *two postcards* you've written. Once she vents at length:

> If you think that your behavior was unknown to me you are sadly mistaken . . . you were drinking far too much for a young girl . . . think about your future and what you will tell your husband about your past . . .

Dad is always the same, opening with sweet words that almost immediately give way to expressions of concern. Most of the time he dictates these letters to his secretary, Evelyn Flagg.

> I have told you how much your letters and postcards mean to us and especially to me. I cherish these communications and would not want to inhibit you at all in writing to me. At the same time, I must comment critically on your very careless and completely unsatisfactory spelling . . .

Sometimes he mentions her weight, saying in one letter that he is "particularly pleased" to hear she is subsisting mainly on fruit and water. He sends her suggestions for papers she might want to write in college:

> William James: often called the father of Pragmatism . . .
> Ortega y Gasset: Spanish philosopher and political thinker who wrote *The Revolt of the Masses* . . .
> Marcus Aurelius: Roman emperor who spent 40 years of his life fighting against barbarian invaders . . .

And he warns her about the kind of left-wing professors who seemed bent on promoting revolution, frequently including articles with typed or handwritten commentary. That November he's especially concerned with the "youth revolution" and tries to make Jennifer see how similar it is to Nazism and Fascism, with its chanting of empty slogans and hostility to different points of view. In one letter, he includes copies of three speeches by Richard Nixon, underlined at the pertinent spots and accompanied by a letter that gives a hint about his current feelings on Vietnam:

I think it is worth remembering that a million and a half Vietnamese fled from North Vietnam in and about 1954 to avoid living under their communist Masters. Half a million were killed or imprisoned by the communists after their takeover in North Vietnam. In other words, the South Vietnamese have resisted and struggled for their freedom for the last fourteen years. They have also put their trust in American support and we have encouraged them to do so during these years. There is no doubt at all that these people who have fought side by side with us and with our encouragement would be massacred in the tens of thousands if the communists were to take over. I have not been able to understand the "morality" of young people in the U.S. who seem so willing to stand by callously and personally contribute in effect to a human tragedy and disaster of such proportions ...

A few months later, Spiro Agnew comes to Seoul for a testy meeting on American troop reductions. As part of the new "Nixon Doctrine," which calls for America's Asian allies to take on more of the burden of defending themselves, the president wants to withdraw twenty thousand soldiers. After the *Pueblo* incident and the North Korean attempt to assassinate him, Park finds this withdrawal so threatening that he's authorized the Korean CIA to fight it with bribes to American congressmen. One after another, they fly out to Seoul for the lavish junket treatment, visits that will later figure in Senate investigations during the scandal known as Koreagate.

For my father, all this must summon up more memories of the terrible years in Vietnam. There's an even more direct echo when he sees the man who arrives at Agnew's elbow. In 1963, John "Mike" Dunn was Ambassador Lodge's right-hand man. I'm pretty sure he was the man who rushed down the hall to turn my father around as he tried to walk out of the embassy without saying goodbye. When the trip is over, Dunn asks him to ride out to the airport in his car, evidently hoping for a chance to put the past behind them.

My father refuses.

That winter my hair grows past my ears and I grow taller than my father. At school, my grades are Cs and Ds in almost every subject. But I'm a whiz in English, where I've skipped ahead two years and have already finished a correspondence course in freshman college English, getting an A that does not noticeably increase my humility.

At home, Dad corrects my grammar: "My teacher and *I*." "I went *farther*." "According to *whom*?"

And I keep making the same mistakes.

But mostly they leave me alone. When I get "caught after curfew" and have to spend the night out, they don't complain. At lunchtime, when I ring the buzzer and ask the cook for a grilled cheese sandwich and a beer, my mother sees the beer on the tray and just shakes her head. They're a bit more worried about my constant camping trips, especially after the Korean militia mistakes three backpacking schoolboys as North Korean infiltrators and shoots them dead. "We are very reluctant to let him go camping at all," Mom tells Jennifer in a letter, "but it is hard to keep him home. We just hope he survives."

Dad peppers me with homilies:

"Don't run with the herd."

"Don't live in the golden ghetto."

"Compassion is the greatest virtue."

But they feel like criticism, like something sharp between my shoulder blades.

Mom's letters to Jennifer suggest that she's worried about him.

> Daddy was terribly hurt that you addressed part of the letter "to mother only." I mean really hurt. . . . You must indicate some interest in his long and informative letters to you, and about the articles he sends.

In November, she hits an unusual pitch of emotion:

> I hesitate to tell you this, and I'm sure your father wouldn't want it known, but when we read your last letter we were both in tears. We miss you so much.

And Dad keeps trying to connect with me. He tells me about being a young soldier on leave and going to the Jefferson Monument, where he first saw that great quote about swearing eternal hostility to all forms of tyranny. He talks about John Stuart Mill and the rights of man and gives me a copy of *On Liberty* to read, says it's one of the foundation documents of liberal society. He tells me that "economic freedom is the beginning of all freedom." And sometimes he repeats his favorite lines in a comical way, raising a finger and an eyebrow and pronouncing his words with exaggerated crispness. "Don't run with the herd, son." Then I know that he's mocking himself a little, trying to lighten the mood.

At Thanksgiving, he takes me on a trip up the Korean coast. We even spend a night in a Korean "yogwan" guesthouse, which takes Dad back to the days of roughing it and gives me a chance to show off how much I've learned about Korea. When we get back he dictates a letter to Jennifer:

> Jocko is continuing to read a lot and expanding his knowledge of classical as well as modern music. He hopes to be able to graduate early in order to get on to college as soon as possible. I sympathize with his aspiration. He is developing fine as a person and steadily maturing.

I write too.

> Dad and I finally got off on that trip, which was very beautiful—fishing villages of straw and whitewashed mud, mountains, sandy or rocky seas, clear green water, rice paddies. You should see it.

By now I've decided that he's very alone, that my mother isn't what he needs, isn't deep enough or something. Look at the way she's thrown herself into the social season with fanatic intensity. One time she goes to four parties in a single day. She just wants to be cheerful and chatter about flowers. She's like a character in Ionesco, ignoring the rhinoceros in the room. When I find a copy of *Everything You Always Wanted to Know About Sex* in her lingerie drawer, it seems like evidence of sadness. I'm convinced they don't have sex, which I blame on her.

And Dad tells me about Philip Wylie and his book attacking true believers, biting out the phrase in that precise bitter way of his. Don't be one of those *true believers* who know the *Truth* with a capital *T*, son. He gives me a copy of *Generation of Vipers*, gives me *Mr. Sammler's Planet* and *Zorba the Greek* and *The Last Temptation of Christ*. He gives me the paperback of *Exodus*, the Leon Uris novel of heroic Zionist struggle. I read them all.

"Don't live in the golden ghetto," he says.

"Don't run with the herd," he says.

And when my reading tastes turn toward the darker moderns, he orders the perfect books for me and leaves them on my bed: *Waiting for Godot, The Trial, The Notebooks of Albert Camus.* These books change my life, but he never mentions them, and neither do I.

At college, Jennifer is alternating between boys and classes. Then Dad says he can't afford to bring her home for Christmas and she decides to spend the holiday in Georgia with her new boyfriend. Mom gets furious and says it's bad form in every way. This Jeff fellow seems to have no consideration for her family or her reputation.

Dad writes to her too, this time by hand. He says that her relationship with Jeff is her business and he's going to leave it up to her, but as a father he should at least express his opinion. He doesn't want to see her locked into marriage or some kind of social outcast.

> You should not flaunt social convention flagrantly or needlessly. Please think this over very carefully, darling. You're still only eighteen years old and you might go a little more slowly . . .

Then he offers her money for the trip home.

She doesn't respond.

In January, Mom writes a blazing letter:

> Aside from the letter asking permission to spend winter term off-campus—and oh yes, a Christmas card—we have heard nothing from you for so long we are in despair. I simply cannot understand why you have been so thoughtless and inconsiderate. Remember your father has had one heart attack already . . .

A few days later, they get an envelope with a Georgia postmark. This time, Dad is the one who responds. "Your letter from Augusta caused your mother great pain. I found her in tears after she had read it. I must add that it brought me no happiness too . . ."

This is when things start to get really crappy. One Saturday, I'm in the sunroom searching his copy of *Lolita* for the good parts. Dad comes in and scowls. "Put on some shoes."

"Why?"

"It's vulgar to sit in the living room with bare feet."

"What's wrong with my feet?"

"Just put on some shoes, please."

"But look at these feet. They're perfect feet. If these feet were carved by Michelangelo, you'd admire them."

"I'm simply asking you to put on a pair of goddamn shoes."

Another time, brazenly climbing the stairs with a glass of beer, I get in an argument with him and tell him not to be so paranoid. It's just an off-hand remark but it turns his face red. *Do you even know what that word means? That's a clinical term! It's a term psychiatrists use!* Of course I have no idea the personal history he has with this term, how it must summon Vietnam and all the snotty reporters who attacked Diem for being too paranoid just before his enemies killed him. But I'm frozen on the stairs and he's barking at me from below, as angry as I've ever seen him. Finally I lose it and hurl my beer against the wall.

In his eyes, I see shock and recognition.

Mother isn't much help, watching our fights with quiet disapproval. She's the master of the little dig you barely notice: "Are you going out in that?"

One night we're talking about June and she says, as if it's obvious to everyone, "She's too good for you."

It doesn't occur to me that she might be unhappy too.

Then June leaves for Hawaii, transferred along with her parents. I drop acid and wander the city, hiking up the mountain called Namsan and through the ragpicker's tarpaper village on the garbage dump near Yonsai University. The Han River is frozen over. A bitter wind blows down from Siberia. Down by the Tongdaemun Market, there's a street where the shacks have dirt floors and girls sit in the windows.

Around this time, a CIA man named Peer de Silva sends my father a copy of a *Playboy* article called "The Vietnamization of America" with a note attached. "Thought you would appreciate knowing that you remain unforgotten by your good friend Halberstam."

The story begins with a flashback to Saigon:

> It was 1962 and the Ngo Dinh Diem regime was at the
> height (if that word can be used) of its powers. The Viet

Cong were stealing the country away at night out in the provinces; but in Saigon, which was all that mattered in that feudal society, Diem and his family controlled all. He won elections by a comforting 99 percent. His photo was everywhere; his name was in the national anthem. He controlled almost every seat in the assembly. He owned the Vietnamese press. The American ambassador is his messenger boy; a four-star American general believed his every word.

To my father, this must seem like the most vindictive kind of fiction. "Controlled all?" "Messenger boy?" "Believed his every word?" It's absurd, a cartoon. Halberstam's larger point is that America is looking more and more like Diem's Vietnam, her cities filled with rats and racial strife and cops beating protesters while men like Spiro Agnew and Nelson Rockefeller gave "stupid Rotary club speeches," which is why Halberstam is growing his hair longer and looking at all the nice young businessmen and wondering what they think of the war.

And for the perfect symbol of the unthinking Fascist bastards behind it all, he chooses my father:

> I did not think of J.R. as being a representative of a democracy. He was a private man, responsible to no constituency. Later, I was to think of him as being more representative of America than I wanted, in that he held power, manipulated it, had great money to spend—all virtually unchecked by the public eye.

He goes on and on, twisting the old knife: J.R. gathered intelligence that was mostly lies and ran covert ops that never worked and "bristled over the problems of working for a democracy," going into periodic tirades about the impossibility of working under a free press. Once he even unloaded to Bill Truehart, "one of the few high-ranking Americans to leave Vietnam with his integrity intact," and Truehart had to remind him that if we controlled the press we'd end up the same as the Communists. To my Thomas-Jefferson-and-John-Stuart-Mill-worshipping father, who defended the reporters in his cables and never said a mean word about any of them, this must have seemed the lowest blow of all.

By coincidence, I choose that moment to engage him on a debate about communism. We are just sitting down to the dinner table and one of the maids is offering me the serving plate and I use the big silver fork and spoon to take some food. You're always talking about how great those kibbutzes are in Israel, I say. But aren't they basically just communes? Isn't that what "kibbutz" means? And what's so wrong with sharing stuff, anyway? All your better hippies live on communes, don't they? Not like the greedy capitalist Rockefellers who—

And Dad explodes, telling me I know *nothing* about the Rockefellers and all the fine civic things they have done to grace our country. Communism is *a totalitarian system of social control that suppresses all freedom*, a system in which self-indulgent brats like me would be the first to go!

He throws down his napkin and storms up the stairs.

Then comes the day Tom and I are walking through the Yongsan gate and one of the MPs calls out, "Hey, get those girls' ID cards." I turn and flip him the bird and we walk real fast up the sidewalk but there they are, coming after us. One grabs me by the arm and I jerk loose and he comes at me with fists swinging. Next thing I know they're dragging us to the guard shack, where they cuff us and throw us in the Jeep and take us to a little jail and tell us we've been charged with assault.

Assault? They assaulted us!

Tom's very quiet. His father comes first and they leave without a word.

Hours later, Dad comes to get me. Driving home in the back of his black diplomatic car, the chauffeur at the wheel, I try to explain. Think about it, Dad. Would two fifteen-year-old kids really attack a couple of huge MPs? MPs armed with guns and clubs? Use your logical faculties. *They're the ones who should be charged with assault!*

He won't even listen. I've embarrassed him, embarrassed the family. I've forgotten that he has a position to maintain in the community. I'm going to have to apologize to General Michaelis. But the minute we get home I bang down the driveway and out the gate, grab a cab to Itaewon and rent a cheap hotel room. My friends come to give me moral support and a reefer.

The next day, not knowing what else to do, I walk down the hill to the high school and go to my classes. When the bell rings at the end of the day, Dad's black car is waiting outside. He drives me to the Eighth Army headquarters and we wait stiffly in the anteroom until Michaelis is ready. Dad doesn't care if this is the most grudging apology in father-son history. He just wants me to go through the motions and be a hypocrite.

The minute we get home, I head down the driveway.

"Where are you going?"

"Out!"

Sometime soon afterward, I convince a pack of stoned teenagers to huddle in a stairwell and read my favorite play:

FATHER UBU: SSSShit!

MOTHER UBU: Well, that's a fine way to talk, Father Ubu.
What a pigheaded ass you are!

FATHER UBU: I don't know what keeps me from bouncing your
head off the wall, Mother Ubu!

Ah, the consolations of literature!

Then it's spring and Jennifer comes back from her first year of college. She made the dean's list and Dad is ridiculously proud of her. I get her high on the balcony and roll her around in a patio chair until she's dizzy, then tell her my new philosophy: there's a whole new world out there that isn't uptight and repressed and the only thing we can do is ignore our crazy parents and live! *Kick out the jams!*

A few weeks later, around the time the *New York Times* starts publishing the Pentagon Papers—the vast cache of secret Vietnam documents leaked by Daniel Ellsberg—I'm packing for a trip through Thailand and India to Israel. The Israeli ambassador suggested that a month in a kibbutz would be good for me and Dad thinks it's a great idea. It'll give me a sense of discipline, a feeling for the real world. I'm set to go in just a few days.

But when Dad comes home from work that night, he comes into my room with the most somber and disappointed look on his face. "A note arrived on my desk today from military intelligence," he says. "It said you were 'a known user of LSD.'"

He doesn't yell at me. He's too alarmed to be angry. Maybe he's freaked out by the thought that his own agency funded research on how to use LSD to control minds. One thing's for sure, the trip is off. He can't ask Yahuda Horam to sponsor a drug user. And it's very distressing that a note like this would come to him through another intelligence service, which makes it clear that my behavior is widely known and certainly well known to the Korean CIA, *people he has to work with every day.* How many times has he told me that I must always conduct myself as *an ambassador of my country*, that he has a *position to maintain in the community* . . .

We end up going to an Army psychiatrist. Jennifer comes too, in her usual role as moderator. But Dad and I each try to make the psychiatrist see how terrible and wrongheaded the other is and it turns into another fight and finally I say that I just want to get away and live my own life. Maybe I can get into college early, like Jennifer did.

The psychiatrist seems to think this is a pretty good idea.

But he's only sixteen, Mom says.

I'll be seventeen in October, I point out.

Dad says he thinks I'm too young. And getting into college isn't realistic. Jennifer graduated. I barely made it through tenth grade. But I keep pushing and the psychiatrist takes my side and finally Dad says fine—if you can get into college, you can go.

34

●

HAWAII, KOREA, HAWAII, KOREA (1971–73)

Hawaii is jet fumes and hibiscus and I'm walking along a road with my backpack heading for June's house. The earth is bright red and the banyan trees drip roots. June lives in a little green ranch house north of Pearl Harbor and boy, is she surprised to see me. She says she'll always love me but she's not sure she wants to be together anymore.

That August, courtesy of sheer bluster and good SAT scores, I start classes at a small private college just across the mountains from Waikiki. I live at the YMCA. My father sends me letters:

> In sending you a check for $300 to enable you to buy a motorcycle, I emphasize that I am doing so on the understanding that you will not operate the motorcycle until you have been properly insured.

When I respond with the same kind of cordial distance, he says he'd like to hear a bit more about what I'm actually doing and thinking. He doesn't mention the big news from home—a group of Korean criminals killed their guards and escaped from a camp on a remote island where the Korean military was training them to slip across the border to kill Kim Il Sung, making it all the way to downtown Seoul before the cops stopped them in a blaze of gunfire and grenades. "Everyone is fine here. Did you receive the check for $100 which I sent you through the mail from Dulles Airport?"

Jennifer passes through on her way back to college and we drop acid, climbing through the astonishing jungle near the Pali gap. A few weeks later, Mom asks her for some intelligence. "Is he doing alright? How about his living conditions, the drug bit? The motorcycle? The college? Please send me a letter and let me have your views."

When she finds out that I'm riding my motorcycle without insurance, she flies out to get me organized. As usual, she's very helpful and frets about all the details but leaves me feeling a bit harassed, as if nothing I do will ever be sufficiently proper and correct. It's a relief when I spend that last afternoon with her and have a goodbye dinner. I drink a beer and leave before dark.

When I wake up, Mom is standing by my bed. You were in a motorcycle accident, she says.

"Was it my fault?"

"No," she says. "Someone hit you."

Then I pass out again and wake up a few days later in a hospital room, my leg up in traction and stitches running from ankle to waist. I have a triple compound fracture and a dislocated hip and a concussion. Mom tells me again that it wasn't my fault, some lady in a station wagon was making a left turn against the right of way. They reached Dad on the military line and he called her and she rushed to the hospital. They wanted to cut off my leg but a heroic doctor named Smith arrived and asked her permission to save it.

I talk to Dad on the telephone and he finds me cheerful and clear headed.

But it's touch and go for the next month. Writing to her brother, Mom says I still could lose the leg if the blood flow falters. And the sad thing is that all my teachers say I was doing very well in school—got a ninety-seven on my first exam.

In his lucid moments—which are not too often yet—he worries about having to drop out of college. He wants to study but he is in so much pain that they keep him well sedated.

For the next few weeks, they wait for my fever to go down and then open me up again, trying to get all the bones set right. None of the doctors have seen three such bad breaks in one bone, so they have to put my case before the hospital board before they make a decision on how to proceed.

Every day at noon, Mom puts in a call to Korea, where my father is waiting at his desk. Even when the morphine has me almost unconscious, I manage to drag myself free and seem perfectly rational until the phone call is over and I drop back into my daze.

Outside my window, cats prowl the dumpster. I spend hours watching them.

When I'm feeling better, I write to Jennifer:

> I'm fine. It wasn't my fault. Next summer, come and stay for a while and we can go to Korea together. There's always a few pounds of pot around (we deal).

For the next two months, I read *Playboy* and *Growing Up Absurd* and push the morphine button constantly. "Not yet," the nurse always says. But one afternoon when the morphine is good, I confront Mom about the old family joke about me speaking Greek and Jennifer having to translate for us. I wonder if she has any idea how pathetic that really is? We didn't even speak the same language.

The next day, she doesn't come.

June visits one day and again the bitter true words flow. It's scary how cruel I can be. She doesn't come back.

Late that year, Park Chung Hee declares a state of emergency in Korea and starts arresting thousands of student protesters. He cracks down on the press and freezes wages. He also starts developing plans to build nuclear weapons and to give a massive push to Korea's six largest industries. This is a direct response to the American troop withdrawal and the ongoing disaster in Vietnam, where South Korea has three hundred thousand troops. If America can no longer be a trustworthy ally, Korea will have to go its own way.

Under pressure for fresh intelligence, my father goes back to brooding about Vietnam. That January, he underlines the salient points of a *Reader's Digest* article refuting the "Big Lies" about the war. It's not a civil war but a war of aggression by North Vietnam. It's unfair to dwell on a relative handful of baby-killing Americans when terrorism is the official policy of the North Vietnamese. Far from devastating the country with defoliation, Americans had actually helped to build Vietnam into the most highly developed country in Southeast Asia.

Back home for the holidays, with hair to my shoulders and a concentration-camp body—130 pounds on a 6 foot frame—I spend my days sleeping and my nights lurching around on my crutches with my face hidden in the fur of my parka, searching for a good time. When I do talk to my father, his abstractions always seem to come out sounding like lectures. His gloomy global concerns seem like the very essence of the uptight adult world. His melancholy scares me.

Judging by Mom's report to Jennifer, I'm in a real daze by the time the break is over: "Jocko left for Hawaii without any money. He left all the green $ and the check for school on his window sill. Typical of him—so absent minded."

Back in Hawaii, I take on a heavy class load and drift from place to place. In March, I write to say I'm living in a garage, please send money. Two weeks later, I want to drop out for a few years. A month after that, I announce plans to stay at a Zen Buddhist temple in Japan. Once, my parents call up the college to try to find out where I am.

But in his letters, Dad gives no hint of distress: "Despite the state of so-called Emergency decreed by President Park, things are quiet in Korea, at least in terms of national security."

That summer, the *Washington Post* breaks the news about a burglary at the Watergate Hotel. The next day, my father's old friend and colleague Howard Hunt makes the papers when the police find his name in the address book of one of the Watergate burglars. My parents don't say a word about it.

Jennifer made the dean's list and I made straight A's, so we're on reprieve. Nobody wants to start a fight, so we try to be polite and stay out of each other's way. But when my cousin Renata comes out for a visit, she notices that both my mother and my father are hitting the cocktails hard every night. My father seems remote to her, absorbed in his mysterious work. He's on edge with me, frequently slipping into annoyance. As Renata and Jennifer devote themselves to a summer of cheering up the troops, often in the skimpiest possible clothing, he becomes oddly prudish. When General Michaelis wonders whether to let the Eighth Army movie theaters show *M*A*S*H*, he argues against it. The scene when they collapse the tent around the nurses' showers is *very* disrespectful to the ladies, he says. He continues to object after all the military generals sign off, which even my mother thinks a bit silly.

But soon I'm back in Hawaii, where I move into a shack without electricity and plunge into a book list that ranges from William Blake to Erik Erikson to Junichiro Tanazaki. Writing to my father, I rail against the educational system and the "artificial" work ethic and tell him to read *Death at an Early Age* by Jonathan Kozol. "In your last letter you said something about the hedonistic argument for achievement, but the school system is based on the idea that you must *force* children to learn . . ."

When my father gets this letter, he's in the middle of another historic crisis. With just one day's notice to Ambassador Philip Habib, Park Chung Hee declares martial law and fires the National Assembly. Habib is furious with my father for giving him no advance warning. Then Park arrests his opponents and unleashes the Korean CIA, an episode later detailed by historian Don Oberdorfer: "In a brutal procedure known as the Korean barbecue, some opponents were strung up by their wrists and ankles and spread-eagled over a flame in KCIA torture chambers; others were subjected to water torture by repeated dunking or the forcing of water down their throats."

I am, as usual, completely oblivious to all of this, caught up in school and drugs and general hippie madness. There's a girl who lives in a cave and a hermit who built himself a little hut exactly the size of one person lying down, like a coffin with a roof. There's a hitchhiker who tells me the CIA killed Kennedy and when I try to argue with him, repeating something my father said about how impossible secrets are to keep, says he's got the secret Warren Commission report to prove it. There are brilliant young musicians who play with a grace and power beyond my grasp. There's a girlfriend who dumps me on orders of her military dad.

To get me to respond to her practical questions, my mother has to send me a check list:

1. Did you receive your air cargo: ____
2. If yes, what did it cost to have it delivered: ____
3. Are you returning home: a. for Christmas?____ b. for good? ____
4. Have you been to see Dr. Smith? ____
5. If yes, what does he say about the pin in your leg? ____

And somehow it all leads to a guy who says he can get me a thousand tabs of acid for five hundred bucks, which would bring in about five grand in Korea and spread a lot of joy and when you consider how crazy everything is and also John Stuart Mill's famous dictum that *economic freedom is the beginning of all freedom*, it does seem quite reasonable to raise the five hundred bucks by getting fifty tabs up front and walking through the International Marketplace muttering "Acid." Although it would probably have been smarter not to actually be on acid at the time.

At the juvenile lockup, I hand my wallet and belt through the slot. When the guy behind the window reaches forward with a hook, I realize he's missing a hand. I spend a dull week hanging in the juvie courtyard reading *Catcher in the Rye,* the only book they have.

Finally a lawyer recognizes me from the accident and contacts my parents. Many years later, Evelyn Flagg paints that little scene for me. How Dad hangs up the phone and immediately calls her into the office. "I want to dictate a cable," he says. She takes out her pad and sits down.

"Effective today, I resign my post as . . ."

As soon as she realizes what he's doing, she interrupts him. "You can't do this!"

"I've got to."

Like many of the people who work for him, Evelyn is very fond of my father. She thinks he's a very decent and thoughtful man, and they have a running joke about his habit of bumming cigarettes. So she's willing to take the risk of talking back to him. "Wait a minute, Jocko! I was in Germany for many years and I saw lots of people who had kids with problems! And the agency never expected them to resign! These things resolve themselves."

He tries to argue with her, but she just tears up the sheet.

"You can't do that," he says.

"Yes, I can. I'm not going to type it."

A few days later, a pair of Hawaiian cops escort me to the airport in handcuffs. When I arrive in Korea, Dad doesn't say much. In his Christmas letter to Jennifer, he doesn't even give a hint of what happened:

> Jocko arrived unexpectedly from Honolulu on the evening of Christmas day. He's in good health, has put on weight, still limps but not as badly as before. He has now finished his first college year.

Mom keeps my secret too, though she vents a little annoyance by saying I forgot my tape recorder and Christmas gifts and driver's license too. "Other than that, he looks good."

The atmosphere improves slightly when my grades come in—three As in literature and a B in Man's Search for Meaning. But like my father's abstractions, they seem to exist in a separate and artificial world. In my real world there's a crazy new edge to things that seems so much more serious and real—last week my friend Adrienne jumped out of a running taxi, just for the hell of it. And Danielle needs an abortion so I steal the old man's Red Label to sell on the black market, taking her home in a cab afterward and chatting with her mother in the same phony how-nice-to-meet-you way I chat with my own mother. And Mike takes me to somebody's hooch and does the whole deal with a needle and spoon and piece of cotton and *jesusfuckingchrist* this room is perfect and golden, a drop of honey oozing through the universe. And one time we wait till after curfew and roll the old man's black car down the driveway to see how close to the DMZ his diplomatic license plates will get us. Driving through the deserted city, we get through about four checkpoints before the South Korean soldiers lose all patience and order us back. And let's not forget Patty, who tells me she got raped by her father *and* her grandfather. And Mark, who goes nuts on acid and tries to convince us that Jesus is coming to Cheju Island and we should all march down to meet him. And back home in my bed I want to write it all down but an hour goes by and I've only written one word and another hour goes by and finally I go down to the kitchen and fill a water glass with gin so I can slow down and get some sleep. In the morning, I read what it took me so long to write:

continuum.

No wonder Dad starts to fret that I'm going to end up "on the beach." I try to mollify him with plans to apply to fancy colleges like Columbia and Harvard. I date a sweet girl who is innocent and studious. But once again the trend seems to be against us. As the Watergate scandal gets tangled up in the CIA, Nixon fires Dick Helms and replaces him with an outsider named James Schlesinger. Then Howard Hunt pleads guilty to charges of burglary and conspiracy and begins a lengthy prison term. Then Schlesinger gathers the Clandestine Service officers in a mass meeting and announces that it's time for a change, the older agents have to make way for young blood. Less publicly, he orders an internal investigation into the agency's most controversial actions, a collection of explosive secrets known to spies and historians as the "Crown Jewels." Winding down to retirement in this poisoned atmosphere, Dad starts to feel weary and a bit bitter. In a note to Jennifer, he makes a rare dig against his new chief with a remark about his age—only forty-three—that packs volumes of generational and professional resentments into a single aside. And right at that moment, in a weird echo of Madame Nhu's edicts against dancing the twist, Korea passes a series of decency laws against "super miniskirts" and long hair.

Dad dictates one of his oddly abstracted reports to Jennifer:

> Korea is quiet under the new order. President Park has assumed complete control of the country so firmly that no one feels he can resist. I don't expect that there will be trouble in the foreseeable future.

Boy, is he wrong. And the first sign of the trouble to come is right here in his own family. Blowing off our parents' warnings about the shoot-to-kill orders, my friends and I pile into a train going to the east coast and drop some acid and make our way to a beach, leaving our backpacks at a nice little clearing in the woods and running *banzai* for the water. But as soon as we clear the first dune we see two barricades of barbed wire guarded by an army pillbox. The soldiers tell us we can't swim here, a boatload of North Koreans came down not long ago and they're on double alert. So we grumble and bitch and slump back to camp and then Mike starts digging through the packs and throwing clothes around and screaming: *We got ripped off! Those fucking soldiers fucking ripped us off! All our blue jeans are missing!* And before we can stop him, he's tearing back out of camp toward the pillbox. By the time we catch up, he's already inside and the soldiers are in an uproar. A few minutes later they haul him out by the elbows and the sergeant starts barking out orders in Korean.

Putting our language skills together, we figure out that he's ordering us to move our camp here, next to the pillbox, for our own protection. Then the soldiers fan out to find the thieves. Apparently we have challenged the honor of South Korea.

So we spend the afternoon tripping in the shadow of a pillbox, watched by hostile Korean soldiers who can't stop making the usual crack about our long hair. *Nam ja? Yo ja?* Man? Woman? And after it gets dark, the soldiers come back with a couple of young Korean guys and take them into the pillbox and then we hear screaming. It goes on for a while. Peter approaches the pillbox door and the soldiers tell him to sit the fuck back down, this is not a game.

Mike gets the shivers that won't stop.

Then two of the soldiers go out into the night. Half an hour later they come back with a bundle and hand it to the captain, who drops it at our feet with a grunt of triumph and contempt:

Our jeans.

Those last months, the stiff invitation cards come in sheaves: the Director of the Korean CIA and his wife request the honor, the Ambassador of Thailand requests the honor, the British Ambassador and Mrs. French request the pleasure, the Netherlands Chargé d'Affaires and Mrs. Verkade request . . . Mom attends as many as three lunches, teas and dinners every day, Dad a different *gombai* party every night. Unbeknownst to me, he is also reaching out to save me, writing secret letters to my sister asking her to do anything she can to get me into college. By now she's a junior at USC, where she transferred after her sophmore year. So try there first, he tells her. He's willing to pay full tuition so maybe they'll overlook my lack of a high-school diploma. And try Whittier College, where an old professor named Dr. Albert Upton might be able to help. And please go at this "promptly and vigorously."

Jennifer offers to let me stay with her in Los Angeles and Mom sends back a grateful note: "Jocko & Dad are on very bad terms—more so than usual. When you wrote that you would like to have him, he was so happy."

And she actually pulls it off, charming the USC admissions people into admitting me for a trial semester. Then the folks leave and I decide to stick around and have one last Korean adventure on my own, moving up to a farmer's house at the end of a dirt road on the outskirts of the city. There's a path up the mountain and a Buddhist temple carved into the cliff and a Peace Corps guy who stands out by the stream cutting soap bubbles with his kendo sword, who loans me *The Brothers Karamazov* and *Thus Spake Zarathustra*. I go from Dostoyevsky's tortured meditations on parricide to "Man is a rope stretched between the beast and the superman." Sometimes I stay up all night and climb up to the temple at dawn, dreaming of the future ahead. I'll prove them all wrong! I'll *become* somebody.

One day, Donald Gregg comes out to visit me. He's my dad's replacement. Looking back it's pretty comical, a CIA station chief and future ambassador to Korea dragging his black car all the way up a dirt road into the mountains to eyeball a co-worker's hippie son. But Dad was worried and asked him to check me out.

Late in July, I write my parents a letter that starts off with a burst of anger. So USC wants to bill me as a "tentative" sophomore until I finish a language requirement? To hell with that! "I was sick of categorized superficial requirements by the tenth grade and I'm not going through them again. I will study only what I want to study and I will not study a language unless I feel like it."

Then I pause to cool my temper, switching into a description of the farmer's house.

> It's Korean style, with arching tile roofs that shelter at least a hundred trilling birds. Mornings, I like to climb the mountain to the temple and sit there in the cool serenity eagle-high above Seoul. From there I can see the River Han stretch like a silver belt from Walker Hill to Kimpo, and the mountains that inspired so many oriental painters to draw craggy misty peaks.

See, Dad, it's no golden ghetto. And the rent is only ten bucks a month!

Did you get *Gestalt Therapy*? Are you using it? In a lighter vein, *In and Out the Garbage Pail* by Fritz Perls also is excellent. I am reading Nietzsche.

I close with "much love," underlined twice. And a final piece of advice: Be happy, Dad.

35

●

MEXICO

There is a photograph of my father's last official visit to the CIA. He's wearing a blue suit and his black horn rims, his bald head tanned from weekend golf. Bill Colby holds out his second distinguished intelligence award and shakes his hand. Behind him, there's an American flag. He looks lean and strong.

But afterwards, when he stops at a colleague's house to say goodbye to a group of old friends, they're struck by how sad he seems, sunk down into himself like a man who feels his life is over. And later my mother tells me that he refused to accept the award from Schlesinger, putting the ceremony off until Colby took over the Agency—one last stubborn moment of principle.

A few weeks after the ceremony, they move into a modest apartment in Guadalajara, a location they chose almost at random because my mother stopped there on the way to Korea and fell for a beautiful golf course. It's cheap and remote and foreign, that's the main thing. Dad will play golf and pick up his fifth language.

In his letters to me, he's upbeat and a bit sentimental. He says my efforts to stay in touch mean a lot to him and he's glad I've landed in L.A. with "wheels and a girlfriend." He's especially pleased that I'm trying to read Kant. "Don't be at all discouraged if he seems totally baffling because he has baffled 90 percent of the people who have tried to read him."

But Mom's letters give glimpses of his real mood. "Your father is finding retirement difficult and as a result he is difficult," she writes one day in August. A month later, she asks us not to come down over the next break. "Your father is not in very good condition. He is pretty damn nervous, difficult, edgy, etc. He is really going through hell."

He tries to keep busy, studying Spanish five hours a day and reading *The Valley of the Dolls* in translation. He visits Mexico City and Patzquaro. But this is the year of Nixon's fight to keep his tapes secret, when he fires Archibald Cox and abolishes the Office of the Special Prosecutor, when the Attorney General resigns and the FBI seals his office to preserve evidence, and once again my father is stuck on the sidelines. Watching his old post from a distance, he sees more chaos. In August of 1973, barely a month after he left Korea, Lee Hu Rak's KCIA agents kidnap an opposition politician named Kim Dae Jung and come very close to killing him, stopping only when Phil Habib rushes to the Blue House to protest. During the massive student protests that follow, KCIA agents torture and kill a dissident university professor. From the CIA station Donald Gregg sends a personal message to Park Chung Hee saying that he doesn't see how he can work with Lee Hu Rak anymore. That's all it takes. Park abruptly fires the KCIA chief, a startlingly responsive decision that raises questions about what results my father might have achieved if he'd brought similar pressures to bear.

By November, Dad is deep in a brooding funk.

> Jennifer, I'm afraid you and I made a mistake on Nixon, understandable at the time and with McGovern as the alternative. I supported him as long as I thought it right to do so, but with the two missing tapes and the late disclosure that they are missing, I feel that I have to give up on him. Our country seems to be going through a terrible crisis at the worst possible time internationally and the exit from this is hard to perceive even dimly.

He's also convinced that he doesn't have much longer to live. His father died when he was fifty. He's already had one heart attack himself.

But Mom stays chipper, sending us travel tips with hidden warning signs: don't wear jeans because the Mexicans don't like hippies and watch out for the restaurants with signs saying they don't serve hippies and oh, one more thing—"Jocko would be advised to pin his hair up."

Fortunately, I end the semester with straight A's. My philosophy professor is especially pleased, praising my first big paper and suggesting I switch my major. So when Jennifer and I go home for the holidays, Dad greets me with a tender green shoot of respect. From the Guadalajara Airport we drive through jumbled streets and diesel fumes to a broad avenue called Chapultapec, where there are fountains and trees and bougainvillea everywhere. Their new apartment is in a small building on a tree-lined side street. It's weird seeing our familiar furniture in yet another new location, doubly weird to see it in a small apartment.

That night, Jennifer and I make our appearances at cocktail hour and give our reports, just like we're kids again. She had a great year too, straight A's and lead roles in all the school plays. Dad nods seriously, asking questions and looking for openings where he can give us advice.

We all want to get along. But it feels clumsy, like wearing clothes that don't fit. And I can't help resenting him a little, since his new respect is based on the superficiality of good grades. And he seems diminished and tentative and the respect feels tangled in that too.

We all end up pretty drunk.

Over that holiday, I probably spend more time with him than I have my entire life. After his morning Spanish class we go for long walks. He shows me the statue of the Niños Heroes and tells me the story of the brave military school cadets who held off the American troops until they were slaughtered, amused to hear himself denouncing American imperialism. He shows me the market and the Orozco Museum and the Diego Rivera murals and tells me the story of Nelson Rockefeller hiring Rivera to paint the lobby of Rockefeller Center, how Rivera put Lenin in the picture and Rockefeller had it painted over. "He'd hired a damn Communist!" He tells me that economic freedom is the beginning of all freedom and his favorite quote is "I swear eternal hostility to all forms of tyranny over the mind of man." He tells me that the only real aristocracy is the aristocracy of the mind. He tells me that his father was six foot two and had blue eyes and "rufus" hair.

They have an active social life. One of their new friends is a flamboyant gay man with a penthouse apartment and a circle of horsey Mexican friends. Another grew up in Imperial China, the daughter of missionaries. Another retired from the post office and loves to play chess. Dad takes me to meet all of them, telling me I could learn a lot from people with so much experience of life. But there's an autumnal feeling leeching out of him and one night after a lot of drinking he snaps and there's a horrible fight. The next afternoon my mother cries. "I should have left him years ago! I just don't want to live this way!" In an instant my sympathies switch from my misunderstood father to my long-suffering mother, saddled with a miserable husband. We're standing in the kitchen and I notice that everything is blue, all the tiles and walls and even the wooden cabinets.

His friends write to him, pushing alternatives. Si Freidin suggests a think-tank job, a Korean friend named Margaret Cho writes on Blue House stationery to suggest reviving some of his connections with powerful businessmen like Dynamite Kim. But he shrugs it off, worried about the expense of life in Washington. When we remind him that he has a good pension and plenty of money in the bank, he says that my mother wouldn't be able to afford servants. And capitalizing on his Korean connections would be unseemly.

Then it's the last night of our vacation and we're all trying to drink our way across the gulf. I pull out my guitar and start playing a song by the Rolling Stones. My sister joins me in the chorus:

> Don't you know I'm a 2000 man,
> And my kids they just don't understand me at all.
> Oh Daddy, proud of your planet.
> Oh Mummy, proud of your sun . . .

Dad usually winces when I try to play, but this time he thanks me again and again. "That was wonderful," he says. *"Wonderful."*

When we finally stumble off to bed, equally drunk, he squeezes my shoulder. "Thank you for the song, son."

When we get home, Dad writes to say how much they enjoyed our visit. In a separate letter, Mom tells us how sad he was when we left. He said it would be the end of him if anything happened to us, and she believes it. And thanks for overlooking that horrible drunken scene on our last night. "It is incredible to me. Dad seems to have forgotten it all. He really does when he drinks like that, I've learned."

A few days later, Dad writes again, this time to Jennifer. He's worried because she talked about being depressed and he wants to tell her that he lost hope in college too. This time he gives details he never gave before:

> I can remember exactly when it happened. I was sitting in my French class, taught by one of my favorite teachers. When she asked me a question and I tried to answer, my voice, without any prior warning nor prior symptoms, croaked unintelligibly just like a frog's. I was unable to go on and felt myself covered with embarrassment and humiliation.

That was the beginning of a steady decline, he tells her. He even developed "ideas of reference," the feeling that people were talking about him. He became neurasthenic, lacking the energy or will to do "even the important things." And late at night, sitting in the hills and looking down on the lights of the city, he contemplated suicide. His teachers and friends had to pitch in to "keep him afloat." Finally he switched to UC Berkeley and made a conscious effort to focus on things outside himself, on art and nature, movies and liquor and the opposite sex. So he really means it when he says that he doesn't care if she's rich or famous, that he just wants her to be happy. Please believe that. Don't forget, his mother supported him till he was thirty. And get a thorough checkup and go to the university psychiatrist. "Meanwhile, at a humble level, please don't overdiet, be sure you have a sufficient protein intake, and try not to worry too much about the future . . ."

Slowly, time passes. My grandmother comes down from Michigan to live with them, which helps give a sense of stability to the household. Quietly knitting an endless afghan or playing cribbage with my mother, she's a calming presence. My father finds her particularly soothing. In his letters to me, he continues to recommend books, George Orwell above all, but also *The Razor's Edge* and *The Gulag Archipelago* and *The Rebel*. And *Pedro Paramo* and *A Hundred Years of Solitude* and *Herzog* and *Humboldt's Gift.* "The universe gets bigger with every book," he tells me. He cautions me against arrogance and quotes Ralph Waldo Emerson: "I can learn from any man." He types out sections of Steinbeck's diaries.

In a drama class, I write a play about an overbearing father who destroys his son with criticism. When my professor says he wants to stage it, I send a copy to my father and he begins to muse on our past.

> I was interested in your reference to the "shortcut" and the "easy way." I remember specifically the first time I used this expression with you. I had asked you to mow the back lawn and when I went out to see how you were doing I found that you had attached the automatic mower with a long rope to an upright pole . . .

But he doesn't take offense. Tentatively, he begins to encourage my ambitions: "Since I'm now fully convinced you are a writer—I was only partly convinced before—I've been thinking about writers and writing, and these are the grand conclusions I've come up with . . ."

A list of Important Themes follows, larded with quotes from famous novelists.

But through the ups and downs and family concerns there is one constant. By spring he says he's convinced that Nixon should be impeached. Clearly he ordered John Dean to pay Howard Hunt 120,000 thousand dollars to keep quiet "in the explicitly stated context of blackmail." That winter, investigative reporter Seymour Hersch follows up his famous exposé of the My Lai Massacre with a front-page story in the *New York Times* revealing the CIA's domestic spying on anti-war protesters and radicals, secret reports with titles like "Restless Youth." In the wake of this new scandal, the Agency fires Jim Angleton and Senator Frank Church convenes an investigation and Dad writes to express his distress. He's known this organization for more than thirty years and has always believed and continues to believe that it's "largely composed of intelligent, well-educated, decent-minded, service-to-country oriented people." Of course there have been errors, but that's to be expected of any big organization.

But he doesn't seem ruffled. He gives more emphasis to his New Year's resolution:

> When Jocko was here, he noticed I was getting out of shape. Too little exercise, too much tequila. I've begun regular exercise again and have pretty well given up on the cocktail

hour routine, struggling to cut down or give up smoking. Happiness is a flat belly.

Before long he's working out an hour and a half a day and walking, and he gets that flat belly. He starts teaching an English class. One day, he's moved by a conversation with a student about the Kantian idea that we should never treat people as means but always as ends, which he considers one of the founding concepts of Western civilization.

He sends me more encouraging words. "We've read your letter several times, and I must say that it seems to me you've made a quantum leap in sophistication, maturity of writing style, and in yourself."

Then the Church Committee calls him to testify about the Diem coup and a couple of Church's aides come down to interview him, along with a reporter from the *Washington Post*. By the time he flies up to Washington to testify he's depressed and withdrawn again. In the space of a few months, he's plunged from deeply meaningful discussions about the categorical imperative to being investigated by his own government for overthrowing a foreign leader. He stays in a spare bedroom at George Mistowt's house and spends a few days browsing CIA archives to refresh his memory, then goes to the committee room and gives his secret testimony.

Later that year, there's another dark reminder. Back in Athens, one of his favorite young officers was a young Harvard graduate named Dick Welch. Welch was cultured and cultivated and spoke fluent Greek. He used to play chess with a Russian diplomat every week, each trying to recruit the other. Eventually Welch became chief of station and moved into our house. That December, in the same driveway where I used to play, Greek terrorists gunned him down.

Like many in the CIA, my father blames Welch's death on a former CIA man named Philip Agee. When he tells me that Agee quit the agency and published the names of many undercover CIA officers in a tell-all book called *Inside the Company*, he spits out his name with a bitterness so intense it's alarming.

But that's about all he says. And he never mentions his Church Committee testimony either. The first time I hear about it is when I'm interviewing Mistowt almost twenty-five years later.

Instead he gives me fatherly advice, CIA style:

When you were here for Christmas, son, I told you that you had an enemy who stalks you day and night—your liberty, your privileges, your hopes and your individuality. The least you can do is to familiarize yourself with the enemy and with the possibilities of defense and counter measures. Bertram Wolfe's *Three Who Made a Revolution* would be a start.

Then it's my senior year of college and I have spent the last two years soaking up the fever dreams of Borges and Beckett, Bulgakov and Marquez. I have a special taste for the morbid alcoholics like Malcolm Lowry and Eugene O'Neill. Dostoevsky is my favorite above all, because his characters suffer so beautifully. I am constantly scribbling the most pretentious horseshit in my notebook: "Freedom means in the existentialist sense the right to be displaced, while the noble savage is the essence of being placed."

My major theme is drunken despair, which recurs like a tic in every story and journal entry:

Hands shaking and shoulders aching from my last binge, sitting in front of the TV in a mindless state, I stagger the liquor stores I go to. I see people walking about and wonder if they have a secret shame and how it is possible for them to act naturally.

Now it's Christmas. In the room next door, my grandmother is dying of heart disease. Dad and I drink and watch *Gunga Din* on the channel that carries English movies with Spanish commercials. When the movie ends, Dad philosophizes. "Life is filled with two things, tragedy and mystery—so we need comedy."

I write it down in my notebook.

"People need kindness, courage, comedy and love," he says.

I write it down.

"Desperation is intrinsic. Death is in this house and we need to laugh."

I write it down.

He talks about the one president he knew personally. "Nixon had courage, but lacked integrity and love. He had one third of the aristocratic virtues."

I write it down.

He talks about his own death in a foregone lamenting tone that has become a kind of family joke, since he's been predicting another heart attack for two decades now. "My one wish is that I will die bravely, with one final ironic laugh on my lips."

When it's my turn, I tell him about my study of Nietzsche and why I decided he was wrong, because even if Christianity is a "slave morality" you still have to be kind to people. Which pleases him, because he has always told me that compassion is the most important virtue. It's one of the Asian traditions and that's something he thinks made my rebellion harder to handle because we were living in Asia, with its traditions of respect for elders. But I have finally become everything he ever wanted me to be. "You've joined the only international aristocracy I know—the aristocracy of the mind," he says. "Now I can die satisfied."

And I tell him that my old anger is resolved too and he smiles his sweet self-mocking smile and answers with arch elaboration. "I'm glad that you've given this to me before I pass into the other world."

"Which is an illusion," I say.

"Yes, but it is still a very warm and comfortable illusion."

What the hell are we talking about? Who knows? By now we're both completely trashed. The point is, he hugs me and I thank him, and for a moment we bridge the awful gap. "Thank you too," he says. "For everything."

But all the sweet reconciliation rings a little hollow because I'm all revved up and I can't sleep. My head is swimming with meaning and poetry and death and it's like I walked into the novel I've been obsessing on all semester, *Under the Volcano*: "The room shaking with demonic orchestras, the snatches of fearful sleep, the voices outside the window . . ." It's my nervous breakdown, right on schedule, just like Dad's when he was in college. And I have to write everything down and also meet the *bruja* who comes to cast out the evil spirits by rubbing the egg all over me and here I am in my <u>temporary</u> room among the maps and bookshelves and the <u>oriental</u> rugs writing it all down—make a note, the <u>alarm clock</u> is broken. Get it? The <u>alarm clock!</u> Do you see how every single word absolutely *bulges* and *throbs* with meaning. Consider <u>accident</u>. Consider <u>shock</u>. Write it down! There's no time to waste! "*Everything* is symbolic—maps, whiskey, lamps, clocks, books, <u>pens</u>, encyclopedias—the *Encyclopedia Britannica!*"

And there's Dad at the door. "Son, you've got to get some sleep," he says.

"Yeah Dad, I will."

"Otherwise you're going to have a nervous breakdown."

"Maybe that's the best thing."

"Yes, but you've got to get some sleep."

I set the alarm for four A.M. There's so much to do!

> DEC. 1: I'm strangely exhausted. I almost passed out just now. Too dizzy. Will write later.
>
> DEC. 1: I am trying (desperately) to sleep, but my blankets keep falling <u>to the right</u>, and must constantly, annoyingly, <u>pull them back.</u>"
>
> DEC. 2: Stones *can* fly. We're standing on one that does.
>
> DEC. 2: To the magician, everything is symbolic. So ALL LETTERS SHOULD BE CAPITALIZED.

And Dad says, "Go back to Greek drama and you will find that the fundamental theme is fatal accident." He suggests I read *Appointment in Samarra.*

I write it down.

Jennifer comes in the room and tells me to turn off the light and go to sleep. I write that down too.

Dad reads the latest version of my play about the overbearing father, a drunken ambassador wrestling with the end of his career. He says nothing about the filial conflict but makes helpful suggestions about some of the subsidiary characters. Maybe the girlfriend shouldn't speak pidgin English. Maybe the wife should be more desperate.

Then he brings up an interesting factual quibble. "No ambassador, unless he had gone crazy, would fail to go to his retirement ceremony. He must show the discipline of his thirty years of obligations, however bitter that may be for him."

He also thinks I should soften the ending just a touch. "Give a little glimmer of hope. That is what the audience will cling to, because the final courage of life is what they need. If the characters end up merely defeated, then people say life is a turd. They need a little glimmer of hope."

But just as everything is warm and glowing between us, I make the mistake of quoting James Joyce's line about making *Ulysses* so difficult, scholars would spend their whole lives trying to understand him. I'm just joking with him, doing my usual devil's advocate routine, but suddenly he explodes. "Any man who says worship me and spend your lives understanding me is a tyrant and a bastard—an *egomaniacal* tyrant and a son of a bitch. That's the cheapest remark I ever heard. It's like adoring a god. Fuck him! A writer communicates to the human community, not to a select band of aristocrats. The Greeks didn't do that. Goethe didn't do that. They *spoke to the human community*."

On my list of true favorites, Joyce is way down in Donald Barthelme territory. But I don't want to give in so I just grunt and nod.

"Think of Shakespeare. He wrote half for the drunk in the second row and half for the intellectual. Do you want to write for your impotent, bloodless clique of critics? Write honestly for the human community!"

Okay Dad, whatever you say.

When the dark mood passes, he tells me a story about the old days when he came back from World War II and went back to Whittier to look up Dr. Upton. Upton launched into a long monologue on the theme of Basic English, insisting with fixed intensity that if everyone in the world would learn five hundred English words there would be no more wars. He didn't ask Dad a single question about what he'd been through. "I left a sad young man," Dad says.

I beg him to tell me these stories, to write them down, to send me letters. "I know nothing about you!"

He shakes his head. "You know I can't do that, son. I swore an oath of secrecy."

A few days later my grandmother rallies and I go back to school. But now I'm revved up so fast, I'm freaking out all my friends and teachers. The head of the Department of Slavic Languages and Literature comes to my apartment to give me a pep talk about Russian Poets: Their Nervous Collapses and Hard-Fought Recoveries. To get to sleep, I have to numb myself with booze. While working at a halfway house for mental patients, where the staff sometimes confuses me for an inmate, I scribble increasingly gnomic *pronunciamentos*: "Put women on the boat and you'll forget about the white whale . . . Jack off the volcano and it erupts . . ."

Even my father writes to say he's worried about me. He wants me to remember his warning that if I kept on at this "excessively intense pace," sooner or later the results would be inevitable. This time he doesn't mention his own breakdown, perhaps because the parallels are too disturbing.

A week later, he writes again:

> The experiences you describe, Russian revolutionary thought, the literary decadents, are all phases of romanticism characteristic mainly of youth but not only of that age. Once again, I feel compelled to prescribe as balance and antidote a great pivotal book, a classic in theory and criticism, Irving Babbit's *Rousseau and Romanticism.*

Then he surprises me. I'll never know if it's because of my extreme distress or because the time is right or if all I had to do was ask, but finally he's ready to give me some of what I need. He sits down at the Hepplewhite dining table and takes out a sheet of thin writing paper. "Well, to begin the saga at the beginning, with origins, my mother's family came over from Russia . . ."

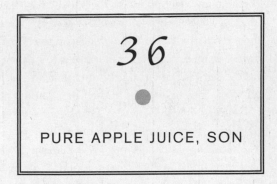

36

PURE APPLE JUICE, SON

And so the years drift by. My sister and I graduate from college and head out into America, trying out cities and lovers and jobs. But over and over, we're drawn back to the jacarandas and bougainvillea of Mexico, to the shifting cast of retired Americans and the new rooms filled with the old furniture. Jennifer and Mom stay up late drinking and playing cards, Dad gets up early and goes for walks that last two and three and even four hours. He teaches English and studies Spanish. We go out to the country club and play golf.

The booze ebbs and flows.

In 1978, his doctor gives him six months to live unless he quits smoking and drinking. When I arrive, he's officially drinking his last drink. I join him and he slips into one of the romantic moods that seem to be coming over him down here in his Mexican retirement, telling me how he wants me to dispose of his ashes after he dies. I must go to the poor fisherman's village of Barra de Navidad and rent the cheapest hotel room in the town, then drink a bottle of cheap tequila and hire an old fisherman to carry me beyond the breakwater to the ocean. I must wait till I get to open water to break the seal. It's a reprise of the eager young scholar drowning his sorrows in Byron and Shelley, touching and a little scary.

After a week of this, my hands shake so badly in the morning that when I grab my legs to stop them, the shaking spreads through my whole body.

So I go back to the States, join a twelve-step program and send him encouraging descriptions of the sober life. I pass on my favorite sobriety motto, "You're only as sick as your secrets." He sends back letters that are tender and supportive, telling me again that I've finally become the son he's always wanted and sending me checks to help me in my "good, mad scramble for self-realization." He admits that his lousiest fault is that when he's hurt, he tends to distance himself in a "shroud of silence." And sometimes he's surprisingly vulnerable: "Thanks for letting me know you're keeping my letters. For a moment I became discouraged at the lack of response."

When my sister falls into a bad marriage and calls him weeping, saying that her husband just threw a punch at her, Dad rushes right over. Although the husband is apologetic and remorseful and extremely polite, Dad is furious. "If you ever lay a hand on my daughter, you'll regret it," he says. "Remember what my position was at the CIA."

He tells Jennifer he'll wait while she packs. She's never seen him so angry.

But in daily life he's the same old Dad, as abstract and distant as ever. When I join him on his walks, he tells me about the history of every statue we pass and discourses on the Stoic philosophers and gives me his version of career advice. "If you're going to be an intellectual, be an honest intellectual. Dishonesty is a form of tyranny."

Then he's summoned back to Washington again. This time the Senate is investigating the Koreagate scandal, which starts with reports that a Korean named Tongsun Park bribed members of the U.S. Congress to support pro-Korean legislation. Dad testifies and comes home and doesn't say much about it. Later a man named C. Philip Liechty passes word to the *Washington Post* that the CIA knew about the bribery all along and never sent the information to Washington. Liechty is one of his former case officers. Later still, Liechty tells a writer named David Corn that my father was "a lush who had accepted gifts of women and liquor from his KCIA friends." There's also a passage in the Senate Intelligence Committee's report that raises disturbing questions.

> During the period 1970–71, Park apparently had numerous contacts with the CIA station chief in Seoul, although their recollections differ as to the substance of the relationship. Park considered the station chief to have been a close personal friend and he claims they exchanged considerable substantive information about Korean politics. The station chief recalls their meetings to have been of a purely social nature rather than substantive. There are no records in intelligence files reflecting what transpired between them. Park recalls receiving a case of liquor from the station chief on at least one occasion.

In various mangled versions, this information appears in accounts of the period. When I read it years later, it hits me with a sickening shock. Gifts of liquor and women? That's flat-out absurd. Dad was far too ethical to accept a gift. He kept two liquor cabinets, for God's sake!

Then I learn that the story is much more tangled than it first appears. Fired by the CIA in 1978, Liechty plunged into a bitter divorce and custody battle that became news when a social worker asked his CIA supervisor for a confidential statement about his background. Apparently the supervisor said that Liechty was fired for making threats and failing to obey orders, details that leaked into the local papers. Liechty then filed a defamation suit against the supervisor and made, in the charged atmosphere of a courtroom hallway, the accusations that ended up in Corn's book. "It was a cover-up," he said. "Internal CIA reports in late 1971 and 1972 had the details, but they were never sent to Washington."

The judge gave greater credence to testimony about his large gun collection and frequent threats to "blast" anyone who got in his way, dismissing the case in less than an hour. But when I study the Senate Intelligence Committee report, I see that Liechty got some of it right. Although my father seems to have sent home "extensive and detailed" reports on Park's activities, they didn't get very far. At one point, for example, the American ambassador to Korea seems to have asked him for a briefing on Park. Apparently my father sent CIA headquarters a cover-your-ass memo asking for permission to brief the ambassador and also for permission to reassure him—all the CIA dirt on Park was being forwarded to the appropriate people at the State Department, wasn't it?

The CIA sent back a surprising response: "Headquarters replied that the information in question was not being passed on to State Department officials, and further, that the ambassador not be briefed."

This suggests that some powerful friend was protecting Park from Washington. Why else would they order him to deny a direct request from the ambassador, risking yet another flap between the State Department and the CIA? It also helps to explain Liechty's venom. Like the junior CIA man from Saigon who leaked the story about the September payment to Colonel Tung, he seems to have blamed his boss for a decision that came from upstairs.

But once again, my father never says a word. It's just another strand in his secret nest of sorrows.

As the decade wears down, his anxiety gets sharper. He fastens on some social error and picks at it until we're all bloody. He's obsessed with politeness to strangers. He broods on his completely imaginary money problems, buying cheap Mexican cigarettes and swapping used magazines with his friends, wearing Mexican clothes because the imported stuff is too expensive. When his latest shot at sobriety fails, he drives out to the tequila factory to get his booze wholesale, carrying plastic liter bottles in the backseat and taking me along to absorb this important lesson in frugality. When I suggest something outlandish like buying a pack of imported Marlboros, he makes a joke out of it: *"Ay, caramba!"*

Over and over and over again, he worries about the impossible careers my sister and I have chosen. He asks me if I think I have the talent to be a writer and when I shrug it off, he comes back an hour later as if he never asked it before:

Do you have the talent?

Do you have the talent?

Do you have the talent?

I don't always handle it well. Then I go back home and he sends me letters musing over our disputes. "I don't understand your 'inner rage,' I never did. Maybe that's why you have a creative streak."

But each time I visit, the cycle resumes. There are horrible scenes. One drunken night, he tells my mother she's given him "no love and no happiness for twenty-seven years," then shuffles off in his boxer shorts to the pantry to rattle around the bottles.

"Old age is a shipwreck," he tells me, drunk.

He tells me again another night: "Old age is a shipwreck. Do you know who said that?"

"Charles de Gaulle."

He's always amazed I know, even though he's told me a thousand times.

Finally Mom says maybe it would be better if I just didn't come to visit for a while.

And he worries about my sister's weight. Beauty is "a fundamental asset in the combat of the sexes," he tells her. A man appreciates a brainy woman but if she's overweight at twenty-seven, he fears what "monument of flesh and fat she'll be at forty." And it's not just a female thing, he insists. "I hate my own fat because it's ugly and betrays profound lack of discipline, of control, even of self-respect."

Once, my sister calls me from Mexico and says she's close to suicide. Dad's been telling her that "the average small house now costs sixty thousand dollars" and we'll never be able to afford one. I relate this to my girlfriend and she seems puzzled. "Couldn't you just live in an apartment?" And I have to laugh because she comes from a sane planet.

And always, he returns to politics. By the end of the decade, he's convinced that the balance of power is turning against America, that totalitarianism is gradually spreading and liberty diminishing. The Soviets and their allies seem to know exactly what they want to achieve and how to go about it, while the West remains divided and blind to its long-range needs and interests. He's particularly disgusted with Jimmy Carter. "Russia is mobilizing in Eastern Europe, has blitzkrieged Afghanistan, and is lining up on Pakistan, and President Jimmy says we'll take serious action—if they're not out of Afghanistan by mid-February, we won't go to the Olympics."

By this time, he's become so conservative he insists he was never a democrat. My mother tries to argue with him. Remember when you despised Eisenhower? Your total opposition to Goldwater? But he just shakes his head and lights another cigarette. I point out that Carter's human rights campaign helped stop Argentina from throwing dissidents out of planes. Isn't that a good thing?

"So you think we're the principal representative of decency and liberty and should go around bringing it about at the barrel of a gun? That's what you think? We're the elect? We're the *real good guys?*"

Every now and then, I get a glimpse of his secrets. One Christmas we get drunk and for the first time he tells me—we are sitting in the garden, looking out over the pool—about a telegram from John F. Kennedy that came with a terrible caveat:

...unless you have overriding objections.

It is the biggest regret of his life, he says. He wishes he had resigned instead of obeying. But it came from the President of the United States of America, dammit.

And he remembers the way Ambassador Lodge delivered the order: "peremptorily."

Then he tells me that Ambassador Lodge said he didn't have an airplane available to take President Diem out of the country, basically allowing his murder to take place. And Lodge must have gone to Kennedy for approval. He's never gone this far before, never come so close to open rebellion. Sensing the importance of the moment, I slip out to the bathroom and scribble some notes on an index card. Later I find out that my sister has had similar conversations and also takes notes, storing them for posterity in a plastic bag.

Then it's 1980. Coming down for another Christmas break, I find him, deep in a sullen binge. Mother tells him that other couples aren't like this. Their friends Marie and Elliott are having champagne in front of the fire. He tells her she's had too much vodka. She starts crying. "You didn't even get me a plant for Christmas," she wails.

Later, I ask him why he doesn't make a little effort, buy her some damn flowers or something. It makes him mad. "I don't know Jenny's birthday, I don't know Eleanore's birthday, I only know yours because it's the day after mine. I think Christmas is a lot of commercial crap and the only reason I care about New Year's is because I may not live to see another one."

But women like a little romance, I say. Remember the story about Grandpa buying Grandma a truckload of flowers.

"And he left her without a penny! I gave your mother a nice home!"

A few nights later, he says something to her that she refuses to repeat. She locks herself in her room.

Then he goes cold turkey and his hands shake so badly that I feel sorry for him. With the worst hangover, he goes out to play golf as if nothing was wrong, goes to his book club and gives one of his sober, judicious reports. He's gracious and dignified to everyone, especially the gardener and the maids. He checks on all the old people, listens to their stories, sends over books and magazines.

His mood improves each day. Sitting on the patio one afternoon, he astonishes me with a story about being a young man in Paris. A whore approached him and he turned her down, and she got mad at him. "What, you come to Paris and you will not fuck the women of France?"

He tells me I'm a good son. His biggest regret is not spending more time with us kids when we were little.

Then he lifts a glass of golden liquid to his lips and gives me the most innocent look.

"Pure apple juice, son."

In 1981, he gives up golf and takes up gardening. He's 67 years old. In a letter, he tells me that you can't really enjoy a garden unless you work on it yourself and start to appreciate the beauty and order of nature. "I find myself increasingly content to stay physically behind these walls," he says.

A few months later, he ends a letter with a personal aside: "Haven't touched alcohol in any form for two months. Thought you'd like to know."

That's the year a United States Federal Court strips Otto von Bolschwing of his American citizenship, the first in a series of very public investigations into the relationship between the CIA and the Nazis. Later a CIA study calls this "the most important Nazi war criminal case involving a CIA asset."

Dad never mentions it.

As he gets older, I worry about him getting conned. He's always so determined to think the best of people. He doesn't seem to have any normal adult suspiciousness at all. He never gets jokes that depend on irony or assume a jaded critical perspective on the world, which means he doesn't get most jokes. But then he'll say we have to support vicious dictators because an authoritarian government can evolve but a totalitarian government can only be opposed from the outside, the same logic Ambassador Jeanne Kirkpatrick used to dismiss the murder of those American nuns in

El Salvador. It's as if his brain has two compartments. In one, he's a hard-nosed cold warrior, a man who lived and thrived in the rapids of history. In the other, he's still somehow innocent.

Sometimes I think that the two sides need each other, and that this explains everything.

In 1982, I come down for Christmas in the wake of a romantic disappointment and make a confession: I've begun to suppress my feelings so completely I'm afraid I've killed my capacity for certain emotions.

"That has always been my problem too," he says.

Back home, I begin to write an early version of this book. My mother tells me he hates the idea. But when I send him the first chunk, he sends back a long letter full of praise and factual quibbles. Even to the most critical parts, he responds openly:

> About my being remote and vague. Part of this may have been the result of your strong rebelliousness from an early age . . . we did go trout fishing once together in the Virginia Blue Ridge mountains, and remember the trip we made together from McLean to trout fishing in Maine?

Then I start working as a journalist and he sends me subscriptions to *The New Republic* and a right-wing broadsheet called *Accuracy in Media*. We begin to debate politics. "I'm not in a position to take issue with you regarding Reagan's environmental policies," he says, "but please remember that no totalitarian country has ever been overthrown in time of peace." Communist regimes in Nicaragua or El Salvador would destabilize the region and lead to other civil wars. Once again we'd end up confronting guerrilla terrorism, "a form of warfare most Americans simply do not understand." And leaving aside our own national interests, he hated the thought of any country falling subject to totalitarianism.

> Having lived in totalitarian Germany and in authoritarian South Vietnam and South Korea, I can only say that there are very fundamental differences between authoritarian and totalitarian regimes.

Over the next few years, he writes fewer letters. Periodically, he slips back into the tequila. In 1983, my sister badgers him about his bad habits and he fires off a defiant note. "The fact of the matter is that after two weeks of total abstinence I wanted to bust loose a little bit. After all, I've drunk and smoked for 45 years or more and it's a little hard to break off completely."

But then my sister comes home fat. Not huge, still under two hundred pounds, but she's been depressed and drinking quite a bit herself and she's put on forty pounds. And Dad gets insanely drunk and says things my sister never forgives.

For the next six years, she refuses to come home.

In 1984, Mom writes to both of us and asks us to write to him more—he's depressed again and seems bored. All he does is sit around reading. By that time, I'm drinking again too. Like him, like Jennifer, I seem to have fallen into a pattern of relapse and reform.

Around this time, I ask him about the blood on his hands. I'm thinking in a general sense of Diem and the war. But he looks hurt and puzzled and doesn't answer. Later Mom gets angry at me. "He never killed anyone or ordered anyone to be killed," she says. "You know that."

Then another year goes by and he goes to the doctor for a routine test. The next day Mom calls me and describes the hurried phone calls to the hospital in La Jolla and the rush to the airport. The next day, I fly out to La Jolla and find him flat on his back in a private room, a giant caterpillar of black stitches crawling up his chest.

It was a triple bypass. But within a few days, he's strong enough to walk around the ward, pushing a stainless steel IV rack with one hand. A few days after that, he's already worrying about the other patients. When we pass their open doors and see their gray faces, he sighs and shakes his head.

A few months later, Mom writes to tell us that he's really turned a new leaf. He hasn't been smoking or drinking and he's a lot more active, enjoying life more and even sleeping better. He's a hell of lot more pleasant to live with too.

Over the next few years, his letters taper to a trickle. In one he apologizes that he doesn't have much to say. "Perhaps these years of retirement and the aging process have had their effect."

By this time, he's drinking less and less. Three years after his bypass, Mom writes a letter saying that he's "been drunk only once since you left and that was not a very bad one." He's better when he's busy so she tries to schedule something every other day or so, a lot of chess and poker and visits with friends. It's a real pity he can't play golf anymore, she says.

Jennifer's in Oklahoma, working on a master's degree in geology and secretly getting fatter and fatter.

The next time I see him is when he flies up for my wedding. This time he's sober and I'm drunk.

He's seventy-three and seems much older.

When Jennifer graduates, he writes to say how impressed he is with her thesis. He sends detailed criticisms of all my newspaper and magazine articles. But Mom writes to say that his spine is "a mess," to quote the radiologist. And the smoking and drinking aggravates his whole condition. "The new thing is that the medication he's on now is nullified by alcohol, so he hasn't had a drink in eight days. Hope he keeps it up."

When the Soviet Union collapses and the Berlin Wall comes down in 1989, Jennifer is down in Mexico for a visit. Mom and Dad stay glued to the television, fascinated by every detail, thrilled by the fulfillment of his life's work. But when I talk to him about it a few months later, he seems amused and almost rueful. All that effort, and the damn thing flops over like a cake.

A year later Francis Winters contacts him with a request to come down and interview him for his book on Vietnam, which sends him into another tailspin. The whole thing is still so painful for him after all these years, Mom says. He writes letters to Bill Colby and the CIA attorney and finally decides he doesn't want to be interviewed unless a CIA attorney is present. Then Winters says he's going to show up anyway.

Mom tries to look on the bright side. "After this is over, he will calm down, I hope."

At this point, I don't know much more about Vietnam than what I've learned from Michael Herr and Oliver Stone. The only evidence I have of his past are those letters he wrote me in college, long forgotten and sitting in the bottom of a drawer. Then I come down for another vacation and Mom hands me a box of old letters in yellowed envelopes. "Thought you'd be interested in these," she says.

All the envelopes are addressed to George Chisler, "Dear George," the first letter begins.

I disturb your serenity of thought for only one reason. Leonard tells me that you entertain some thought of coming to Berkeley next year. If it lies within my power at all to encourage you in this project, I urge you to do your very best to come here—Berkeley is one of the most wonderful places in all the world; it has a beautiful campus, beautiful buildings, most lovely surroundings. All the professors here are stimulating and interesting, and some of them are true artists. They have succeeded in reviving my faith in culture, a faith which stands between me and utter despair of life. . . .

37

●

CONVERSATIONS WITH
OLD SPIES

Another five years go by. Now Dad is pretty much sober, if only from exhaustion, the sweet guy from the Chisler letters finally breaking through the realpolitik crust. We get along great. On his eightieth birthday, I fly down to surprise him. But when we go out for a walk now, it's three blocks and he has the route planned so he can sit down a few times along the way. My mother tells me that if I have any questions to ask him I better ask them soon. But I know from experience that he's not one of these guys who takes a vague general question and runs with it, so one day I find myself driving out to Whittier to dig up some fresh intelligence. The orange and lemon groves are long gone, the main road one long strip mall, but down in the basement of his old high school they still have all the yearbooks. Flipping past Richard Nixon, I find a dapper young man with serious eyes and a wry smile—Jack "Chinkface" Richardson. The caption reads like a prophecy: "Never flashy, but always in the thick of the battle, he proved in satisfactory manner to be a very capable guard."

At lunch, cousin Ray tells me that Strelsky means "palace guard." In his seventies now, a big man with a broad face, Ray remembers the time Dad hiked up the Sierras in dress shoes and the time he refused to share his silver doctors at Humphries Lake. He says that great-grandpa Strelsky was a tall blond blue-eyed man who came from somewhere in the Austrian-Hungarian Empire, possibly Monrovia, possibly on the run from some kind of trouble with the royal family, then went to South Africa in the late 1880s to look for diamonds. And he gives a very different take on Uncle Leonard's death. In his version, Leonard went out hunting with a boy named Victor York and a revolver, not a shotgun. When they got home he went down in the basement to clean it and it just went off, a complete accident. Ray rushed over from school and found Leonard stretched out on the floor, my grandmother still trying to revive him.

Soon George Chisler joins us, telling me about meeting my father at the First Baptist Church and the many hours they spent discussing Will Durant's *Mansions of Philosophy* and the time they skipped the high-school prom to see Dracula. He opens his high school yearbook to the page with my father's picture, pointing to the note in my father's handwriting: "George, don't forget unworthy Chinaman. He's not as bad as his nickname sounds."

Chisler has bristly eyebrows and a merry smile. He's wearing blue jeans. Soon we are joking like old friends about my father and his odd ways. When I ask if he was always so damn serious, Chisler nods a bit wistfully. "He was. That's the way he was."

But he doesn't want to talk about the breakdown year or the Berkeley mania. Here's a story that will put your father in a nutshell, he says. Jack was slated to make valedictorian, but there was a girl who really wanted it so he withdrew his name. "That's the kind of person he was. He was a wonderful man. I loved him."

These stories take my father out of the family and put him back into the world, which is oddly pleasing. Next I go to Washington to meet Si Bourgin, who is tall and thin and still enthusiastic about everything, living in an apartment filled with books and art. After settling his wife on a comfortable sofa—she has Alzheimer's—he pulls out his old *Time* magazine files and takes me back to Vienna, sketching out the political situation and important personalities like the old reporter he is. "Jocko was a total professional who ran a very disciplined operation," he says. "But he had an element of sadness and shyness."

Hearing my father's name, Bourgin's wife comes awake for a moment. "He was a decent man," she says.

Then it's lunch with a couple of old spies named Bronson Tweedy and Dave Whipple, both of them age-spotted and bald and full of the merry irony of worldly men. They tell me that Dad was one of the best, a pillar of the Clandestine Service, a tough guy who took strong stands. They remember him drinking lots of coffee and smoking heavily, his "slightly ponderous way of expressing himself." Tweedy remembers his last night in Vienna, dancing to gypsy music at the Monseigneur nightclub and walking home in first light. Later I learn that Tweedy became chief of the Africa division, making the history books for his involvement in the assassination of Patrice Lumumba. Whipple was a station chief in Cambodia. But they don't seem to carry the past the way my father does.

It's the same story when I drive out to Virginia to visit Gordon Mason. Genial, nattily dressed, still wearing a rakish Errol Flynn mustache at eighty, Mason makes me a dandelion salad just the way he used to during the Italian campaign and tells me about Spingarn and the Palazzo Leonetti, the dreamy Italian girls and the time Dad astonished him by springing for lobster in Venice, about Greece and the planes and boats and radio stations and all the agents they sent into Albania and lost. Knowing my father's tendency to downplay all glamour or drama, Mason wants me to understand how important he was. In many ways, he outranked the ambassador. He had more prestige, more contacts, more men and money to command. "You look at him now and you wonder at the power this man held in his lifetime. But he never flaunted it. Unlike some. He was very modest."

But when I get home, I find an urgent e-mail waiting: my mother's had a stroke. I break off my reporting to rush home and Dad meets me at the door, looking dazed and weak. "I'm very glad you're here," he says.

At the hospital I find my mother well enough to come home, though still very sick. When I go to fetch her in the morning, Dad insists on coming along even though he's so frail I have to make him sit down on a bench to wait while I find the room. They're both exhausted by the time we get home, but Mom settles back on her pillows and starts to talk. "Last night I kept thinking about the time your father's plane got shot down."

Her voice is a slurred drawl. One corner of her mouth tugs down.

"It was the night of our New Year's party, and General Don called to tell me. He said Dad was dead. I called Dad's secretary and she said 'You weren't supposed to know about that.'"

Dad stands at the foot of her bed. His hearing is so bad now, I have to repeat everything she says to him at shouting volume: "SHE SAID, YOU WEREN'T SUPPOSED TO KNOW ABOUT THAT."

"They were probably just trying to protect you," he says, his voice tender.

She answers back with the same tenderness. "I went out to the airport to wait for you."

The moment lingers, the terrible memory turning sweet between them. A few days later, hoping to tap the same alchemy, I sit my father down in the library and break out my tape recorder. He tells me that his favorite high school teacher was a woman who taught ancient history and that once, when he was on a trip with the basketball team, they went into a store and stole some candy. He's astonished when I tell him his high-school nickname. "Now that you mention that Chinaman thing, it seems to emerge painfully. I had forgotten it years and years ago." He asks me if any of his friends had anything bad to say and I tell him that one of them said he was "correct but unimaginative." He nods his head. "I accept that. That's probably right."

I tell him Ray doesn't think Leonard committed suicide. "I'm glad to hear that," he says.

He's quite open about the nervous breakdown. He remembers thinking that people were talking about him behind his back, thinking of suicide. He remembers a night when a girl kissed him on the cheek and he responded by taking her for a walk down a "street of beauty," gushing about his mystical enthusiasms. "Well, that girl never went out with me again. I've never forgotten that."

When we get into the CIA, he tells me stories he's never told before, like the time a young Jewish officer in Vienna offered to go into the Soviet sector of Vienna to try to recruit the most active Soviet intelligence officer under the cover of ... what was that organization?

"The Jewish Rescue Committee," Mom says.

I shout it to Dad: "THE JEWISH RESCUE COMMITTEE."

He nods. "So I accepted his proposal and he went in, and he didn't succeed. Then I sent in a cable informing Washington after the fact and headquarters comes back: 'Cease all anti-Soviet activities pending review.' So I sent in a cable saying that I feel that headquarters had lost all confidence in my stewardship of the station and I hereby offer my resignation. And Jim Angleton told them, 'Don't accept it.'"

Why were they so upset?

"They probably felt, 'My God, he's risking this young officer's life,' and that was no kidding. It could very well have led to his death, and I hadn't cleared it with Washington."

Mom reminds Dad that he tried to resign in Greece too, and Allen Dulles wouldn't let him.

"What did she say?"

"YOU TRIED TO RESIGN IN GREECE BUT ALLEN DULLES WOULDN'T LET YOU."

He nods. "She's getting tired but I think she'd like to hear this story," he says, telling me that during the Suez crisis, the Greek foreign minister wouldn't give the United States permission to fly its warplanes over Greece. So he was drinking with a Greek friend and vented his frustration in a careless remark: *"Evangelos Averof is the worst foreign minister in Europe!"* Of course his remark got to Averof within hours. A day or so later, he was summoned to a meeting in Italy with Allen Welsh Dulles, the director of the CIA. That's when he tried to resign. But Dulles just told him not to be so loose with his tongue, ending the meeting with a chuckle. "You know John, Averof isn't really the *worst* foreign minister in Europe."

After my mother goes to rest, he tells me about a time in Vienna when he crossed through the Soviet zone with maps detailing the locations of military forces in Yugoslavia on the backseat. At the checkpoint, the Russians questioned him closely and inspected his car. "But they didn't touch the maps, which would have shown me to be a goddamned spy," he says. "If they had looked at the maps I might not be here talking to you."

"Aren't you supposed to use a hollowed-out cane or something? A secret panel in the car?"

"It was foolish," he admits. "A dumb thing."

I bring up *Mole,* trying to flatter him with his big counterespionage triumph.

"As I recall he approached some American at his car and defected," Dad says.

Wasn't it a big feather in your cap?

"I suppose it was, but it was very accidental," he says. "I don't think we deserved any particular merit."

At least tell me that Joel Roberts was savvy and tough. Tell me he knew every alley in Vienna's Innere Stadt.

"I think it was a little exaggerated."

But he keeps returning to Vietnam. The story Anne Blair told about Lodge cutting his access to the cable traffic? It was all about that cable Taylor sent to General Harkins saying Kennedy really hadn't made up his mind, he says, adding new details to the story he told me long ago. But there was no explosion and no withheld cables, that's the important thing. "As I recall, the ambassador didn't say anything and we left the office."

Another time he calls me into his room, where he's lying on his side under his blue electric blanket. He wants to tell me about the time his name got on that Vietnamese hit list. Right after he learned he was on it, Khiem called. "Have you heard that I had an assassination unit?"

"Yes."

"Would you come over for a cup of tea?"

Knowing that poison was a favorite Vietnamese assassination technique, he made a joke and tried to change the subject. But Khiem persisted and finally he agreed, figuring that at least he'd have his chauffeur and car outside the house and it would be "a little scandalous" for a leading citizen to assassinate him with his chauffeur waiting. "So I went over there, he offered me a cup of tea, and he talked for about two hours at least, almost as a monologue. Very friendly. And we parted on friendly terms."

But what he remembers most is the cable ordering him to play his part in the coup and the phrase that put the responsibility on his shoulders: *unless you have overriding objections*. "I did send a cable toward the end in which I said that Ngo Dinh Nhu had been a very bad influence and that if it were at all possible, he should be removed. When I use an expression like removed, I mean re-assigned or something, I'm not talking about assassinations. But when I look back at that, I regret that cable. I realize it was impossible. It just wasn't possible."

He also remembers the day he tried to walk out of the embassy in Vietnam without saying goodbye to Lodge and an assistant stopped him. He went into Lodge's office, then Lodge shook his hand and told him he was a victim of policy decisions. "You're tough," he said. "You can take it."

And shortly after he returned to Washington, he had a meeting at the White House with Kennedy's national security advisor, McGeorge Bundy.

"I remember saying that I was surprised that Ambassador Lodge had not spent more time in Saigon before making up his mind about Diem. And I still remember McGeorge Bundy saying to me, as we were walking toward the door of his office, 'John, he had already made up his mind even before he left for Saigon.'"

He sounds almost amused.

My mother isn't. "Lodge wouldn't go into the office until 10:30," she says. "In the afternoon he would have lunch and stay home for a nap. Everybody was talking about it."

"Don't forget he had a serious heart condition," Dad says.

That's when Mom takes me to the maid's room and points to the bottom drawer of her file cabinet. Under some commemorative plaques, I find a pile of fat manila envelopes stuffed with a treasure trove of pictures, report cards, letters and photographs. There's an envelope of 8×10 pictures of Dad's crashed plane. There's an envelope bursting with newspaper clippings from the summer of 1963. I flip through the parade of headlines my parents lived: WAR IN THE PAGODAS, VIETNAM: CRISIS OF INDECISION, KENNEDY TO DIEM: WIN PEOPLE OR LOSE WAR AGAINST REDS. And here's the famous turning point, the 150-point headline on the front page of the *Washington Daily News*:

THE CIA MESS IN SOUTH VIETNAM

It's the Richard Starnes story that shot my father out of Saigon. I take it to the library and wait while he reads it. "This is a lot of crap," he says. "I told Lodge about the message from General Taylor as soon as he got to his goddamn office. I'm not conscious of having disobeyed Henry Cabot Lodge."

"I'm sure you didn't," I say. To cheer him up, I read bits of a speech Senator Thomas J. Dodd gave in Congress in February of 1964:

> In May 1961, Richardson briefed me for seven or eight hours, the most detailed, the most balanced, most knowledgeable briefing I have ever been given. But I was even more impressed by Richardson as a man. Indeed, of all the

hundreds of people in the American service whom I have met in the course of my travels, I can recall no one for whom I formed a higher esteem than John Richardson.

Dad grunts and I continue. Recently, Dodd said, an article in the *Washington Daily News* crossed a dangerous line by openly identifying a CIA representative abroad, "thus reducing, if not destroying, his potential usefulness forever." And clearly there was an official source *behind* the article, an officer who was himself guilty of violating rules of secrecy and basic ethics, an officer who should be "identified and dismissed."

And who was that official source?

Dad sees where I'm going. "I can't imagine Ambassador Lodge floating a story like that," he says. "He may have had his failings, but it seems to me way beneath his stature."

Later I find a cache of cryptic notes in my father's handwriting and feel the surge that comes from uncovering a secret—maybe *this* is it, the final clue to the mystery.

Disenchantment with Diem govt. and Diem. Corruption.

"Looking at things in retrospect," he says, "corruption is one of the cultural aspects of all of Asia from the Philippines to China to Japan. And Mexico. And Chicago."

Selection of officers for loyalty not skill.

"When one looks at that in retrospect, was it all that unreasonable that Diem was looking after the survival of his administration and even of his own life and his brother's?"

Oppression.

"Some people have argued there wasn't enough repression."

Toward the end, the notes become more personal. One mentions our abandoned allies, "one of the keenest pangs of defeat." Another questions America's decency. "Nat'l interest—cold blooded. Cut our losses but written in human blood." In the last, under the heading <u>Worst Episode of My CIA Service</u>, he asks: "Why didn't I protest more?"

Then he lists the possible answers:

Machine gunner image—carrying out orders mentality
Highest authority and centralized information and judgment
Excessive modesty
Pension?

Conclusion—lack of sufficient conviction in thesis that Diem was indispensable.

What does it mean, Dad?

"I was probably thinking about that cable that said, 'unless you have *overriding* objections to the decision of the President, you should carry out the coup plans.'" As always, he says the phrase "overriding objections" with a bitter twist.

Excessive modesty?

"I don't have any comment on that."

Pension?

"That was probably a crude self-interest consideration. I suppose self-interest plays a role in most people's decisions."

I seriously doubt it, I tell him.

"Have it your way."

This is when I begin to question this search into my father's professional past. The novelist Don DeLillo once said that he was fascinated with spies because "they represent old mysteries and fascinations, ineffable things. Central intelligence. They're like churches that hold the final secrets." But I've come to be suspicious of the part of me that responds to this kind of thinking, the writerly way of making things fancy with metaphors. Sitting with my father on the cracked leather chairs, my mother's ancient cocker spaniels farting on the faded Persian carpet, I wonder if it isn't just another childish dream, like having superpowers or a magic ring or learning the secret name of God. Maybe the truth is right here on the surface, waiting for me to see it. So I turn the conversation to more personal things. Mom says that when she met him, Dad was fat.

"I think I probably got plump in Greece."

"You were plump when I met you, and he muscled in on a date."

"I didn't hear that," Dad says.

"MOM SAYS YOU MUSCLED IN ON A DATE."

"Oh, you mean with Si Bourgin?"

Mom nods. "And he was dating three other women at the time. Everybody talked about it."

Now this is fresh intelligence. "What did they say, Mom?"

"That he had all these girls."

"So he was a playboy?"

"Yeah, he was a real playboy."

And she tells the old stories, the bachelor party when someone hurled a mustard pot right past the ear of John Erhardt and they smashed glasses on the floor, the balls and the Dior gown and Balenciaga mink and the baby pig he carried all through their wedding reception, mysteriously converted to cold cuts by the time they got back from their honeymoon. "He didn't tell me he was in the CIA until after we got married. I thought he was a real bona fide foreign service officer."

"I was perhaps carrying security rather far," Dad says.

Was he romantic? Did he bring flowers?

"Let's not get too personal about all that stuff," Dad says.

But when she's tired and ready for bed again, Dad gets up and leans forward to give her a kiss.

38

•

MEXICO, 1998

Then Jennifer calls again. "He's been spitting up blood," she says. "He kept it a secret for a few days but yesterday he finally told me. He didn't want me to tell Mom."

And guess what? He's having difficulty breathing too because he's got fucking lung cancer. One big lump and a couple of small ones. It was asymptomatic so the doctors didn't see it. Now they say he won't last another two weeks.

When I get off the plane he looks really bad, the skin tight on his skull. He spits blood into a kidney-shaped dish. My mother and sister tell me he was very grim and despairing two days ago, railing against the drawn-out miserable process of dying. But since they put him on the synthetic morphine he's been fairly cheerful. Now Jennifer asks him if he wants to eat.

"Why eat?"

"Because if you don't, you'll get weaker."

"Maybe I should get weaker," he says.

In the study tonight, we watch sitcoms. Dad's voice has become a whispery tissue. "I remember the old days in Vienna," he says. "Dean was the youngest major in the Army."

Twenty years after he retired, Uncle Dean finally told us he spent the war as the MK-Ultra officer for the 5th Army, an important player in the secret project to decode Nazi cables. Some say it made all the difference. Now he's an ocean away, also dying of cancer.

Dad smiles all the way through *Spin City*. When it's over, the nurse helps him to bed. "It was a good night," he says.

At ten, he wakes up and calls for me. I help him into the study. Jennifer is about to get breakfast but Dad waves her back. "I think she should be in on this," he says. He sits on his little Greek chair waiting. He's hooked up to an oxygen tank, breathing through thin plastic tubes. Every few minutes he spits blood into the kidney-shaped dish, dabbing at his lips with a napkin.

Finally we're all ready and he begins. "I feel we're not making any progress," he says. "When you came down on Monday we felt it was a matter of days but I feel—*I* feel—that this could just go on and on. So I want you to call Mike and talk to him."

What Dad means is that he wants me to talk to Doctor Mike about giving him some kind of suicide shot. He pauses to spit into the dish and I carry it to the bathroom and wash it out, trying not to look at the bloody phlegm. When I come back I ask if he's concerned about my job and convenience, because he really shouldn't worry about that. I'll do what I have to do.

"I suppose I could go off the machine," he says, meaning the oxygen.

I look over at my mother. As it happens, this very morning a friend of hers sent over some real morphine left over from the death of her own husband. I tell Dad about this and say we could always put a batch of it by his bedside if he wants. This may sound terrible but I know it's what he wants and I don't want him to suffer anymore. And when he frowns I try to reassure him, because I know exactly what he's thinking. "It's not like your brother," I say.

"I've always felt bad about my brother's suicide," he says. "I wouldn't want the grandchildren to think their grandfather did that."

"You put up a great fight for eighty-four years, Dad," I say. "It's not like you're taking the easy way out."

My mother and sister are weeping. The maid vacuums in the hallway.

"I know you feel like it's dragging on," I continue, "but the doctors say it'll just be a week or two more. You're not in pain, your brain is still sharp, and Clinton still hasn't been booted out of office. Why not let nature take its course?"

He seems pleased by that. "Just a week or two?" he says.

I nod.

"Then let's let nature take its course."

"And if you start to suffer or just feel you have reached the end of your rope, then know that we do have this alternative," I say. "Talk to me. You don't have to tell Jennifer or Mom. Just come to me."

"Okay then, we'll wait one more week."

Then the phone rings and my mother picks it up. "Sid, I can't talk right now," she says.

The next morning Dad is in so much pain the nurse gives him an extra synmorphine tab against doctor's orders and straps a heating pad to his shoulder with a bathrobe belt. Two hours later the pain is back and he sits there wincing while she massages the back of his neck. She avoids the left side, where there's now a huge goiter-like gland swollen with cancer. We're all freaking out, so we push her to give him more pain pills and she talks to the doctor who repeats his instructions: one synmorph every four hours and that's it. Then the nurse leaves the room and I slip him three of the real morphine tablets, and they work great. An hour later he's smiling and thanking Mom for letting him have them.

Dad hasn't eaten for three days. The guy who runs the nursing service suggests a synmorph drip, so I get Mike on the phone and he agrees to write the prescription and I make a run to his office to pick it up and they bring over this packet that Dad can carry around with him like a cassette recorder and stick a needle into his belly to start the drip. An hour later Dad slips off to the bathroom and tries to rip it out. I try to convince him to leave it in and he stands there his pants around his ankles and says he just doesn't like it and doesn't want to be hooked up to anything and just doesn't *like* it, dammit. My mother reminds him how he hated the oxygen mask at first, and how he fought the catheter when he needed that last year, and finally he gives in and sits watching *Crossfire*. But as the day goes on, he gets more befuddled and scared. I hate what this is doing to his dignity.

How long will this go on? We have no idea. The guy who runs the nursing service says that even though Dad hasn't eaten for three days he could go on for weeks. With some people it's just like that, death takes forever. As long as they keep getting fluids, they just live off their fat and muscle tissue and slowly waste away.

Meanwhile, Mom and Jenny squabble. Jenny mocks Mom's standing order to the servants to move food from large Tupperware to smaller Tupperware as the portions get smaller. Then Mom snaps at me and I snap back, "Don't snap at me!" She breaks into tears.

Tonight he's bad again. We're trying to watch *Fly Away Home* because he's an old softie and loves the story, which is about a kid who learns to fly a plane so she can lead a flock of motherless geese to Florida for the winter. Something about it gets to him. Things start out okay but as the movie goes on Dad gets more and more confused and out of it and starts insisting we take out the morphine drip. Finally we give in and have the nurse remove it. But then we can't find the Tramadol pills he was taking before and Jennifer goes into a crazy tirade about the nursing service stealing our medicine and leaving Dad helpless against pain. Mom tries to calm her down so she doesn't upset Dad and Jennifer swings a pillow at her. It ends with her fleeing the house in tears—another nightmare. I go out and try to talk to her but she keeps repeating her complaint, which seems much deeper than the actual issue at hand. "They took the drugs," she says. "They had no right to do that."

Then Dad gets better again. He sits in the study with the oxygen tube wrapped from nose to tonsure like Salvador Dalí's mustache and he raises three fingers. "What is that?" he asks.

"It's the morphine, Dad," I say.

This is our new secret code.

Then he starts joking around about being afraid to turn his back on our cat, which has a vicious streak, giving us that goofy old look of mock alarm, a face I now make to my own kids. "You've still got your sense of humor," I say.

He smiles. "Two things, son," he says. "The first is humor and the second is courage. I'd like you to tell the grandchildren."

At 10:30 he says he'll stay up another hour. Jennifer is sitting there with bleary red eyes, my mother's half asleep, I'm hoarse from shouting into Dad's deaf ears. We exchange quick looks of distress.

Dad sees. "I can't do everything," he says. "I can't go to bed just so you guys can go to bed."

An hour later he finally gives up and I sit down next to him on the bed and he points at his stained old golf hat and I put it on his head. He smiles at the nurse, his face in profile so thin and noble. I want to draw him, to take a photo, to keep this moment somehow. And it's funny because I was just thinking about *The Death of Ivan Ilyich* and wondering if the cosmic breakthrough at the end was something true or just something Tolstoy dreamed up to make himself feel better about dying. And here's the old man tiring us all out, staying up long after we're ready for bed. So tenacious.

It ain't cosmic but it's something.

Then Mom comes in and leans down for a kiss. "Long voyage," he says, smiling at her with those bright beady death eyes.

 $\mathbf{D}$ ad wakes up at seven. When I come in he's lying back on his bed with his shirt open and his bony ribcage showing. He's in pain and makes the sign of "the three." The nurse has just given him his morning pills, so I just give him two. He drifts in and out after that. Later he tells me the pills almost "overwhelmed" him. "Don't tell Jenny," he says.

 $\mathbf{A}$ t 3:30 we finish watching *Fly Away Home*, Jennifer and Mom and I all weeping through the last half hour and Dad smiling in perfect Buddhist happiness. When the credits roll I smile at him. "You liked it," I say.

"Loved it," he says.

Then he sits across from me in his slippers and blue plaid pajamas, reading the paper. He doesn't want to take a nap. "At this point I take nothing for granted," he whispers, with a smile.

 $\mathbf{J}$ ennifer says she's sick of this sadness and misery and wants it to be over. I don't blame her. She's been doing this every day for two months. And he's asking over and over if he can take a bath at nine, if it's okay with the nurses, and please is anyone confused because there are two nines—two nines in a day, understand—and he wants to watch the news shows at ten and eleven so he can't take a bath at nine in the morning so please make sure it's okay with the nurses and if anyone is confused there are two nines—

It's exhausting. And then he gets annoyed when we don't understand.

I decide it's time to test his new painkillers. Later I'm in the kitchen gorging myself on potato chips and onion dip when Jennifer comes in and starts whining. "You're going to eat it all!"

I'm not, I tell her. Anyway there's another bag in the pantry. And a thousand more bags down at the Gigante in Plaza Patria. But she won't be mollified. She goes on and on. "You get everything and I get nothing."

Dad goes to the bathroom and sits and sits and nothing happens. It's really hurting him. Not just discomfort but pain and what seems to be a kind of deep biological frustration. Jennifer suggests that this is because he hasn't eaten for four days, so Dad weighs death against constipation and finally decides to drink a protein shake and some prune juice.

The next day he's still constipated. Then he wants to go to the hospital but decides he doesn't want to go to the hospital even more, so he drinks another shake and more prune juice and starts hacking it right back up. Carrying the kidney-shaped dish to the bathroom, I gag and almost vomit myself. I have to admit I'm getting a little annoyed. How dare he try to rally? He's supposed to be dying. And I'm starting to hate that infernal little Frankenstein pacemaker that keeps ticking his heart over and over no matter what the rest of him wants and needs. I can see it under the mottled skin on his chest, hard and round like a hockey puck. Sometimes we joke about passing a magnet over it and putting him out of our misery. Dad nods out, forgets what he's saying, vomits again. Meanwhile the TV news prattles on and on in the background like an evil guest that won't go away.

At around noon Dad says he wants to have another talk, so we gather in the study and he says, pitifully, "My bowels have shut down."

The idiot blathering of CNN continues, distracting him for a moment. "And something else, what else has shut down? My intestines?"

We turn the sound down and try again.

"Your lungs, you said."

"Yes."

Jennifer speaks gently. "It took a few days the last time," she says. "Maybe if you wait . . ."

Then he forgets what he was going to say and we turn the news back on and the dog starts digging in the trash can and my mom starts fretting and my sister says she'll go get the garbage can from the guest house because that one has a lid. "It's not a garbage can, it's a trash can," my mother says. "There's a difference you know."

Dad has been listening. "I wish there was a lid for me," he says.

"That's pretty funny, Dad."

"Do you think there's a lid for me?" he asks.

I raise my hands to the heavens, taking the question for whimsy. But he persists.

"Do you think a doctor would do it?"

"What, Dad?"

He dips his head, his eyes going confidential. "Give me a lid for me."

It's odd how very old people get childlike when they tell a secret. For a second I feel ancient myself, older than he is, and I put my hand gently on his knee. "I don't think a doctor will."

Then he nods so wearily that we try again to convince him to go to bed. But he won't. Never would, never will. Has to finish his prune juice first. And I remember the old days when I'd see him walking to the kitchen at dawn with his tequila glass in hand, and sometimes he dropped it and we would find the bloody footprints later. And now when his head droops with sleep, I try to pry the glass out of his hand without waking him and he jerks back like I'm trying to steal it. I laugh. Finally he drinks it down and I say, "As always, Dad, you drank it to the last drop." And I can't help feeling proud of him.

Mom's in bed. Jennifer and I come in, hovering, needing to talk. It's getting to be so hard on him, I say. It's hard on us too, Mom answers. Which is a sentiment worth honoring, I think. Then she starts to cry and says she didn't think he'd wake up this morning. Maybe we should call the doctor. A doctor put her friend Mary to sleep and woke her up every few days to see if she was still in agony and finally just stopped feeding her through the tube. Maybe Mike would put a lid on him, put him in a deep sleep. Jennifer says we should call the vet and we have another sick laugh and I think: maybe it's up to me now. Maybe I should just do it and spare them the choice. So I go on the Internet and search for the Hemlock Society and discover it's all philosophy. "Where's the fucking how-to section!" I say.

Jennifer laughs. She's looking over my shoulder. "It's ridiculous," she says. "If you search for 'terrorist handbook,' they'll tell you how to make a pipe bomb."

"Maybe we can use a pipe bomb?"

"It might not work," she says. "He's pretty tough."

This morning we thought he was dying. Then he finally succeeded in the bathroom and now he feels *much* better. But so tired he didn't even watch the news. And when he goes to the bathroom again, he asks me to sit with him. Leaning on the edge of the sink, head hanging, he says, very emphatically: "Remember—this—is—lung—cancer." When he's finished I pull up his pants. The pillow-damp hair is stuck wild to his head. But weak as he is, he still insists on washing his hands, leaning over the sink with his elbows on the tiles.

I call home and my youngest daughter says she's fallen in love with a book called *Ella Enchanted*. She loves it so much she took it to a slumber party and read it while the other girls watched the Spice Girls movie. I tell this to Dad. "That makes me very happy," he says. "I couldn't be happier. Tell her I said that."

He's peaceful tonight. Lies quietly, rises only to drink milk or medicine. Asleep at nine. I think the end is coming soon.

In the bathroom he sits on the toilet for twenty minutes. I sit in the chair across from him. The bathroom is all yellow. There's a black ink drawing of a rearing horse on the wall above him. I can tell he's thinking deeply about something and finally he says it. "If—I—need—something, ask—your—mother—first. Because—we—have—the—past. I can't—remember."

I want to be sure I know what he means. "If you need something specific, or anything?"

I have to repeat it a few times before he understands me.

"Anything," he finally says. "Because we have the past."

At three in the morning I wake up to the sound of the nurse pounding on his back. He sits there gasping, head hanging, breathing the oxygen from the tubes. When he recovers, he says, "I can't take this anymore."

The nurse does everything she can to help him. It pisses me off. I point to the oxygen, to the pills. *"No está bien, está malo,"* I say in my mangled Spanish: It's not good, it's bad. *"Necessita morir."*

He needs to die. But I'm afraid to give him the morphine. And I'm afraid not to.

We go to the bathroom. It's the third time, or the fourth, or the fifth. Later it will seem like we spent the whole night hobbling to the bathroom. He leans on us with both arms, bent in half.

At around four he hisses out his frustration, "I—can't—die."

It's not going to end here. I'll keep searching for him in old books and old spies. I'll read *The Story of Philosophy* and catch a distant echo of that earnest young man so fascinated by learning, struggle through *Rousseau and Romanticism* and sigh at his humble tolerance for academic horseshit. In the *Meditations* of Marcus Aurelius I'll find a sense of how hard it is to keep your equilibrium in an awful situation, how a man comes to terms with life in a savage world. As he predicted, I'll find him all over *The Last Puritan*. In *Mole*, I'll find him less in the stories of CIA derring-do than in passing remarks: that punctuality is "a common obsession of operations officers," that "anxiety is the occupational disease of espionage." There I glimpse my father, the anti-Bond.

And I'll interview David Halberstam and Neil Sheehan too. Sheehan will be a friendly, silver-haired man who looks like John Updike, stooped from arthritis. He'll dig through his old card file to find phone numbers for me and say that he thought Dad was naïve, that he thought too well of Diem and didn't understand the hostility of the peasants to the strategic hamlets. "People were always dividing this into the lesser of two evils. When you look at the fact that your Dad worked with the police, who were torturers, or his people worked with the police, your Dad was working on the assumption that this was the lesser of two evils. Communism was the greater evil."

Halberstam will sit for an interview in a grand Manhattan apartment with a curved staircase and tell me that Dad was a decent guy in an unbearable situation, a faithful servant of a flawed policy. He'll tell me that he didn't use our name in the follow-up he wrote to the Starnes exposé, insisting that someone in the head office saw it on the wire and stuck it in. But he won't remember much and will seem strangely oblivious to his own responsibility in Vietnam, blind to the link between his support for the overthrow of Diem and America's deepening involvement in the war. That's the great irony, that he was one of the "best and brightest" too. And when I ask why he was so cruel in that *Playboy* article, he'll shrug. "I was a lot angrier then."

I'll come up with a million theories to explain my father. I decide he's a character out of Hemingway, with the illusion of no illusions, holding onto grace-under-pressure and a cocktail glass. I decide he's the perfect expression of the postwar generation, plating anti-communist steel over small-town innocence and letting out wisps of their bruised longing in drunken parties, in the weary romanticism of Frank Sinatra. They were so eager, so heartbroken, so angry. And we were onto them by the time we were fifteen. Then I decide that's the real reason behind the 1960s, not the war or feminism or civil rights. We burst under the pressure of their sadness.

But there are still so many questions. I'll never understand how people can be so damn sure about what to do in the Middle East or equatorial Africa when I can't even answer the basic family mysteries. Did Uncle Leonard kill himself? Did Grandma Annie develop mental problems? What happened with Dad's "German friend?" And what about that hasty marriage in the squirrel cage? What about the day his new bride spurned Jonathan Swift and dashed his *Pygmalion* fantasy? Why did he sell his big new house? Why did he go to Greece? How much did he know about Frank Wisner's work with the OPC? How did he feel about it? How much did he know about the "Korean barbecue?" What's the real story about his connection with that Nazi, von Bolschwing? How close did he get to the Ustaša? And what about Unit 684, the group of Korean criminals who escaped the camp where they were being trained to slip into North Korea? With my father's long experience with border-crossers, it seems likely he was informed. Maybe he was involved.

It's another thing I'll never know.

The truth isn't hidden, I remind myself. It's right out there in the open. All you have to do is look. But sometimes I come across things that disturb me. Here's a cable from Halberstam telling his editors not to pay attention to the stuff Maggie Higgins was writing. "She spent most of her time interviewing head of CIA now thoroughly discredited . . ." After reading that, I can't sleep. "Thoroughly discredited?" What an arrogant jerk! The fucking guy was sneering at my Dad two days after he landed in Saigon! And somehow it turns into a night panic: I'm a fool, life is slipping by me, I'll never do anything worthwhile! And *how will I pay for retirement*?

It's odd, given how hard I rebelled against him myself. What do I care about Ngo Dinh Diem?

One time my friend Chris—who loved my father like a son—tells me about the time he read an article about the CIA teaching Iranian secret police how to torture and asked him in a shocked voice if he knew about such things. My father gave him a sorrowful look and answered: "Terrible things happen in this world."

I think about that.

And I think about what Gordon Mason told me once on the phone. "Your father had a conscience that nagged him all the time. He didn't have the killer instinct. That hurt him in this business."

A few years ago, reading the newspaper, I came across a letter to a newspaper editor about the CIA. "Secrecy corrupts," the writer said, "absolute secrecy corrupts absolutely." Once again I feel that gorge of irritation. Where do you get off being so self-righteous, Mr. A. Homer Skinner of Marblehead, Mass.? Would you have preferred there were no OSS to fight the Nazis, no attempt to fight the Cold War, no CIA today to dig dirt on international terrorists? We don't do this with Army Intelligence or the NSA or the FBI. But onto the CIA we project all our anxieties about being grown-ups in an ugly world. And it's so easy to point the finger. So easy to sit at your desk and write *critiques*. What's not easy is to choose between the possibility of a global gulag and the lives of thousands of innocent Filipinos or Iranians or Nicaraguans and then live with that choice alone, as my father did.

Then there are the endless questions of Vietnam.

Sometimes my sister and I wonder if we've exaggerated everything, dramatizing our childhood through the scrim of our troubled adult lives. She reminds me of what a great dancer he was, how the women were charmed by him, how much he loved parties. In Korea he competed in the "best legs" contest, and Mom always said he would have won if not for embassy politics. Sometimes I think about the books he started leaving on my bed when I was fifteen and I hated him. I try to picture the romantic Berkeley boy who read poetry and wore that "flowing, multi-colored scarf." I'm so sorry I never met him. And I'll never know if Dad killed him out of shame or if he just held the knife straight while history pushed it in, but I do know this: as time passed, he replaced his doubts with convictions and became so absorbed in his war he forgot that happiness was part of wisdom. And that is a sad, sad thing. And a dangerous thing too, because when you become too sure that life is a tragedy, then little by little you begin to accept tragedy and finally something perverse in you begins to invite it.

One last trip to the bathroom. Even now he won't use the bedpan. We spend a long time in there, just the two of us, sharing a private moment. Then the toilet paper roll is almost empty and that's when he says his last words:

"Another roll."

I get one from the closet and hand it to him.

As the dawn light rises in the window, his breathing starts to change. The agonizing long pauses when you think he's stopped and then a gasp sucking the air back in for one more round. Long pause and gasp, long pause and gasp. It's horrible. There's something monstrous in those sucking gasps at air, something so hungry and automatic, like his self and will are just the creature of this tyrannical little spark of survivalist life that forces him to go on and on and on. Outside the birds are twittering and then the church bells ring as they do every morning here in Mexico, rolling out into the still suspended air. Then Dad calms. His breathing gets softer and shallower breath by breath, with no more gasps or sucks, until he's breathing so peacefully, so gently, just skimming off the thin air at the top of his lungs. I move up and sit on the edge of the bed. Mom and Jenny sit at his feet. The bells are finished and now the garbage trucks rumble by.

The breaths get shorter and shorter and then he just stops.

SOURCE NOTES

This isn't one of those semifictional memoirs where the writer uses imagination to fill the gaps. Except in a few historical scenes where conversations were remembered years later by the participants, every line of dialogue in this book is documented in notes or tape recordings. For greater transparency and to combat the scourge that is footnotes, I've tried wherever possible to cite sources in the text. Where an event or detail is disputed, I've tried to note that as well. The early part of the book is based on personal observation backed by notes, on extended taped interviews with my father and mother, and especially on the letters my father wrote as a student and young intelligence officer. From the CIA, my most crucial interviews were with Gordon Mason, Bill Hood, James Lilley, Lou DeSanti, Sue Darling, Evelyn Flagg and Donald Gregg. I was also informed by conversations and correspondence with Howard Hunt, Gordon Messing, Bronson Tweedy, David Whipple, Marga Meryn, Seymour Russell, Jean Nater and Roz Matthews. Other CIA officers helped anonymously. The civilians who gave critical interviews include David Halberstam, Neil Sheehan, Rufus Phillips, George Chisler, Horace "Tully" Torbert, Ray Zufal and Si Bourgin.

Beyond this, some specific citations are necessary.

Chapter 9:

Arthur Hulbert's quote appears in a book by Ian Sayer and Douglas Botting called *America's Secret Army: The Untold Story of the Counter Intelligence Corps*. I'm also indebted to the *Saturday Evening Post* series by Stephen J. Spingarn, published November 27, 1948.

Chapter 10:

Julius Sagi's war diary can be found in the National Archives, Box 49, Record Group 319.

Chapter 11:

The Carla Costa story comes from interviews with Mason and my father, the National Archives files, Spingarn's *Saturday Evening Post* series and *America's Secret Army*.

Chapter 12:

The quote about "the most productive period in the history of the CIC" is from *America's Secret Army*.

Chapter 13:

The material on Yugoslav exiles came from the National Archives, Record Group 226, Box 46. The material on the Crusaders and Macek comes from Record Group 226, Box 638. My father's name file is there too. General John Magruder's quote appears on a document in Box 518. I'm also indebted to a book by Mark Aarons and John Loftus called *Unholy Trinity: The Vatican, the Nazis, and the Swiss Banks*.

Chapter 14:

Much of the detail on Vienna immediately after the war comes from a military study called *The American Occupation of Vienna: Planning and Early Years*. Interviews with Si Bourgin, Bill Hood and Sue Darling carried on the story. Other details come from the National Archives CIC files. The story of Josef Hutter appears in Box 89, Anna Wukowitz in Box 116, the letter from my father's office about Otto Abrecht von Bolschwing in his name file. John Erhardt's cable can be found in the 1948 Austria volume of the *Foreign Relations of the United States*, the invaluable collection of policy documents put out by the State Department. I drew the conclusion that my father's situation reports stayed calm from the CIA's Vienna intelligence estimate in February 1949, which said "the Kremlin would be reluctant at this time to take the risk of war entailed in a blockade of Vienna." This document can be found in the National Security Archives.

Chapter 16:

Howard Hunt's story about the body at the Vienna morgue is from his book *Undercover: Memoirs of an American Secret Agent*. An essay in a National Archives publication called *U.S. Intelligence and the Nazis* pointed me to Wilhelm Hoettle, whose cables appear in his name file in the National Archives. The Otto von Bolschwing documents can be found in his file. Timothy Naftali's quote appears in another essay in *U.S. Intelligence and the Nazis*. The story about the axe handles came via a CIA agent who asked to remain nameless. The story about the snatched operative is a mixture of Bill Hood and my father.

Chapter 18:

The Pointdexter story appears in Howard Hunt's *Undercover*.

Chapter 19:

For my portrait of Frank Wisner, I'm indebted to Evan Thomas's work in *The Very Best Men*.

Chapter 20:

Joseph B. Smith's memoir is called *Portrait of a Cold Warrior*.

Chapter 21:

Wherever I refer to cables and official memorandums from the Vietnam period, they can be found in the three Vietnam volumes of the *Foreign Relations of the United States* for 1961 to 1963. The agency historian who wrote the history of Vietnam asked me not to use his name, since his history is still classified. William Colby's Vietnam memoir is called *Lost Victory: A Firsthand Account of America's Sixteen-Year Involvement in Vietnam*. My account of Diem leans heavily on Phillip

E. Catton's excellent book, *Diem's Final Failure*. I'm also indebted to George McTurnan Kahin for *Intervention* and Ellen Hammer for the passionate scholarship of *A Death in November: America in Vietnam, 1963*. The Army study on the Special Forces and the Montagnards is called *The Green Berets In Vietnam: 1961–1971*. Prochnau's evocative book is *Once Upon a Distant War: Young War Correspondents and Their Early Vietnam Battles*. Harold Ford's official CIA study is called *CIA and Vietnam Policymakers: Three Episodes 1962–1968*. My father's account of his talk with Nhu about funding the strategic hamlets appears in Francis X. Winters's book, *The Year of the Hare: America in Vietnam, January 25, 1963–February 15, 1964*.

Chapter 22:

David Halberstam's first account of his meeting with my father appears at much greater length in *The Making of a Quagmire*, where he calls my father "a good man, honest and dedicated." Many years later, he revealed his suspicion that my father must have been trying to manipulate him to William Prochnau during their interview for *Once Upon a Distant War*. George Patton's story of the helicopter crash came from Brian Sobel's authorized biography, *The Fighting Pattons*. The CIA study that cited my father's reports on the "steadily deteriorating situation" was *CIA and Vietnam Policymakers: Three Episodes 1962–1968*.

Chapter 23:

The story that the bodies of the Buddhists were examined by a "distinguished physician" appears in Stanley Karnow's *Vietnam: A History*. The version with a "Buddhist doctor" comes from Ellen Hammer's *A Death in November*. The line about "direct, relentless, table-hammering pressure" comes from John Mecklin's memoir, *Mission in Torment: An Intimate Account of the U.S. Role in Vietnam*.

Chapter 24:

Marguerite Higgins included her story about my father and Robert Thompson on the roof of the Caravelle Hotel in *Our Vietnam Nightmare*, the book she based on her contrarian reports from the Diem years. Anne Blair's short and thorough study of Henry Cabot Lodge is called *Lodge in Vietnam: A Patriot Abroad*. Colby's account of the events that followed the Cable of August 24 appears in *Lost Victory*. His reaction to my father's "point of no return" memo appears in Francis X.

Winters's *The Year of the Hare*. The most dramatic version of Lodge's reaction to the "second thoughts" cable appears in William J. Rust's *Kennedy in Vietnam*.

Chapter 26:

The story about Lou Conein handing the Cable of August 24 to my father just as Lodge walked in comes from Anne Blair's *Lodge in Vietnam*. Another version can be found in *The Pentagon Papers*. His description of his decision to place the Saigon station "at the complete disposition of Ambassador Lodge" appeared in a letter to Anne Blair dated November 13, 1990. She didn't quote it in her book, but he retained a copy for his files. Additional details come from William Rust's *Kennedy in Vietnam*.

Chapter 27:

The Australian journalist who tipped my father about the assassination list was Denis Warner. His account of the events appears in his memoir, *Wake Me If There's Trouble*. Colby's account of his visit to Saigon appears in *Lost Victory*. Roger Hilsman's quote about the payment to Tung appears in his memoir, *To Move a Nation: The Politics of Foreign Policy in the Administration of John F. Kennedy*.

Chapter 28:

Hilsman's take on Lodge's attempt to fire my father appeared in *To Move a Nation*, Halberstam's in *The Making of a Quagmire*. The quote about the leak being "the prerogative of the ambassador" comes from *Mission in Torment*. The CIA study that noted Lodge's sharp criticism of my father was the study called *CIA and Vietnam Policymakers: Three Episodes 1962–1968*. The quotes from Lodge's private memoir appear in Anne Blair's *Lodge in Vietnam*.

Chapter 29:

John Mecklin's skepticism about the Starnes article appears in *Mission in Torment*.

Chapter 30:

The account of Diem and Nhu's deaths that includes the briefcase of cash and Conein appears in Hoang Lac's essay in *Prelude to Tragedy*.

Chapter33:

Halberstam's *Playboy* article was published in January 1971.

Chapter 34:

Don Oberdorfer's book is called *The Two Koreas: A Contemporary History*.

Chapter 36:

C. Philip Liechty's quotes come from David Corn's *Blond Ghost: Ted Shackley and the CIA's Crusades*.

ACKNOWLEDGMENTS

Ross Wetzteon was the first editor to show an interest in this material. In 1983, he commissioned an early version of it for the *Village Voice*. After reading the first draft, he gave me a simple piece of advice that stuck with me ever since: "Set the scene."

Ten years later, David Granger asked me to write it for *GQ*. After reading the first draft, he wrote an edit letter that made me rethink everything I was doing as a journalist. "You're using your reporting to avoid judging your father," he said.

Both those stories got killed. But five years later Granger asked me to try again for *Esquire*, where executive editor Mark Warren finally guided it into print. He kept telling me to unleash myself and aim higher. "The story of you and your father is the story of the twentieth century," he said, so many times I started to believe it.

But there are so many guiding angels. Heather Schroder helped me turn the story into a book proposal and sold it to David Hirshey at HarperCollins, where Hirshey and a bright young man named Nick Trautwein encouraged and cautioned and otherwise guided me through three years of writing. Chris and Tanis Furst helped with memories and research and unflagging support. Ali Flint cheered me through some tough spots. In addition to their very useful interviews, George Chisler and Gordon Mason encouraged me and vetted their sections of the book in draft form. James Lilley reviewed the Philippine section. Si Bourgin shared his files. John Taylor patiently steered me through the National Archives. Kathy Potter and Rachel and Julia Richardson lived with me and sustained me in every important way. My sister shared her memories and her diaries and let me put in all the most horrible stuff, in addition to getting me into college and helping me pay my bills for a few years. Without her, truly, the odds of this book being written were slim.

Finally, there's my mother. She had spirit. She carried many burdens graciously. At a terrible time in her life, she spent many hours sitting for interviews and reading drafts and digging up old phone numbers. I miss her.